Blueprint for Reconstruction

Blueprint for Reconstruction

The Rebuilding of our Planet's Urban and Ecological
Infrastructure and Perfection of Life on Earth

Threshold to Meaning Series
Book Three

VINCENT FRANK BEDOGNE

WIPF & STOCK · Eugene, Oregon

BLUEPRINT FOR RECONSTRUCTION
The Rebuilding of our Planet's Urban and Ecological Infrastructure and Perfection of Life on Earth

Threshold to Meaning, Book 3

Wipf & Stock
An Imprint of Wipf and Stock Publishers
199 W. 8th Ave., Suite 3
Eugene, OR 97401
www.wipfandstock.com

ISBN 13: 978-1-55635-926-2

Manufactured in the U.S.A.

The individual seeks self-perfection.
Humanity seeks perfection of life on earth.

Contents

Illustrations

Preface

OUR DESIRES AND OUR dreams may vary. Our road is and must be our own. Yet, beneath the myriad of human expression, the motivation that drives our journey through time is the same. By virtue of our birth, by virtue of our humanity, by virtue of our status in evolution, we each play an essential role in the universe's becoming. It is our nature as individuals to look ahead. It is our character as human beings to move forward, to push the reaches of who we are. We are compelled to advance to greater autonomy and consciousness, to greater personality and substance of character. We are driven to build a better future for ourselves and for our families. Inescapably, our greater sense of self has a practical outcome, one that touches our lives in a down-to-earth way. Our evolution must result in a reshaping of the physical world and in a reinvention of the way we go about our lives in that world.

The present book is the third in a series of three titles that expand on the *evolution of consciousness* view as put forth by the twentieth century scientist and philosopher *Pierre Teilhard de Chardin* and that delve into the impact of this view on our lives and future.

In the first book, *Threshold to Meaning: Book 1, Evolution of Consciousness*, we explored humanity's nature and purpose. We proposed a fundamental mechanism of change, the *creative process*, explained how this mechanism worked, and presented an account of the universe's origin, evolution, and future as logic and reflection tell us it would appear in light of our fundamental unifying principle. A key outcome of this account was the observation that the universe is today undergoing an evolutionary transformation—that the universe is today at a turning point that will impact our lives in the most significant way. This transformation manifests on many levels. By way of the individual and expressed through the human experience, the universe, today, crosses the threshold to meaning.

In the second book, *Threshold to Meaning: Book 2, Economics of Fulfillment*, we narrowed our focus to expand on one level of the evolutionary transformation that we explored in the first book and that is now taking place within us. At present, we live in a world dominated by *scarcity-based economics*: by socialism and capitalism and the values and view of the world they signify. In light of the difficulties that characterize today's global economy—and with the insight into the human experience gleaned by virtue of evolutionary context—our second book began with a simple idea: the notion that contemporary economic practices—those that learn toward socialism and those that lean toward capitalism—are obsolete and that we can invent a better way to conduct our economic affairs. We established this premise and then developed an economic philosophy able to meet our material needs and our needs as evolving beings in an evolving universe: an economics for the human beings we are today becoming—an economics of fulfillment.

As evolution advances and economics of fulfillment systems take hold, we establish the social and economic conditions that allow us to live in almost any environment we imagine and take upon ourselves to create. In the current book, *Threshold to Meaning: Book 3, Blueprint for Reconstruction,* we address the second major level of change brought about by our evolutionary transcendence. We present the vision of the earth's future rural and urban landscape that our understanding of ourselves and where we are headed compels us to formulate. We then draft a plan to bring into reality that which is in our minds. We rise above the polarity between the forces of exploitation and environmentalism that today lock humanity in the abyss of stagnation, collectivity, and central control. We develop the framework to look beyond the kinetics of urban sprawl and incoherent development, beyond the dynamics of environmental dogma and conservation for the sake of conservation—and the police power of government through which today's organizational forces manifest—that at present shape our cities and countryside. We design a physical landscape that addresses the highest calling of our humanity—a landscape able to meet our needs as evolving beings in an evolving universe. We put forth a blueprint for reconstruction of the earth's *urban and ecological infrastructure.*

As the third title in the *Threshold to Meaning* series, the book expands on a key outcome of the evolution of consciousness account of the universe developed in the first book and, in doing so, builds on the ideas developed in the second book. Like the economic philosophy of tomorrow, our building of the cities and countryside of tomorrow is an aspect of humankind's crossing of the threshold to meaning, which is an aspect of the universe's greater evolution. It is possible to read *Blueprint for Reconstruction* without the background of the prior works—and I have done my best to make it accessible as such—but the topic's depth and greater meaning will be lost. Blueprint for reconstruction is not a plan for the future as we think of one today. It is a plan for the future as we, today, think of one in light of our understanding of humanity's status in existence—a design for the world's urban and ecological infrastructure that, as we cross the universe's threshold to meaning, we achieve the wisdom to create.

In part one, *Landscape*, and in chapter 1, *Perfection*, which establishes the first part of the book, we provide the background necessary to develop the ideas that follow. We highlight the evolution of consciousness view and the economics of fulfillment philosophy, trace the evolution of the earth's urban and ecological infrastructure from its origins through the advent of urbanization and agriculture, explore the economic and environmental doctrines that today shape our landscape, and look at the conflict that exists between these doctrines.

In part two, *Blueprint for Tomorrow*, we build on what we have learned in the book's first section and direct our vision into the future. We begin by asking ourselves a simple question: In light of the evolution of consciousness view and what we know about ourselves, what kind of a world do we want to live in? In light of humanity's threshold to meaning and transcendence to economics of fulfillment, and what they tell us about ourselves, what do we want our cities to be like? What do we want our countryside to be like? How do we want to get around and communicate? We then draw a conceptual plan that meets the design objectives our questioning has allowed us to establish.

In part three, *Implementation*, we put our plan to work, or at least we look at how in the future we might do so. In what way will reconstruction begin and how will reconstruction tie in with the economic reformation poised to grip humanity—with the emergence and spread of economics of fulfillment? We tighten our focus to the nuts-and-bolts of implementation, and draft a working drawing for reconstruction. What will our rural highway system be like? What will be the population density of our cities?

How will we meet our energy needs? How will we design for a biosphere that is in evolution and for our planet's natural cycles of climate change?

In part four, *Perfection,* we look beyond the first stages of implementation and explore reconstruction on a global scale. In chapter 11, *Seed of Genesis,* we look at when and in what way urban and ecological transformation will spread across the planet. We project our vision further ahead in time and look at what life will be like in the world we create. Evolution progresses by building on and creatively discarding the old to create the new. To what end will reconstruction lead? To what outcome will perfection of life on earth open the way?

Today, humanity stands at a crossroads in evolution—at the turning point we call the universe's threshold to meaning. As we mark this transcendent moment, we embrace an awareness of our purpose. We comprehend where we have been, understand where we are at, and grasp the destination to which we are headed. With this clarity as to where evolution will sweep the human experience, we have the means to envision the earth we want to live on and to draw a blueprint to create that earth. Free of the conflict inherent in the economic practices and environmental dogma of today, the physical world is ours to mold. The landscape is ours to contour, to nurture, to preserve. Long ago, evolution's forward arrow locked in humankind's advance. In the years ahead, reconstruction of the earth's urban and ecological infrastructure will form the nexus of our evolution. With the intent that comes with an enlightened sense of purpose, humanity will unify in its calling to perfect life on earth.

1

Perfection

O UR OBJECTIVE IN THE upcoming chapters is to envision the rural and urban landscape of tomorrow, draft a conceptual blueprint to construct that landscape, and look at how we will put our plan into place. To achieve this objective, we begin with a point we have spoken about at length in earlier works and must bring to the forefront of our thoughts and discussion. The world as we experience it is a product of our knowledge and understanding. It is an outcome of our notions, concepts, and philosophies. It is a derivative of our worldview.

With this assertion, a brief review of the material covered in the first two books in the *Threshold to Meaning* series is in order. For the most part, we, today, define our universe in physical, or *external,* terms.[1] Inspired by the notion of scarcity and reinforced by the materialistic ideals of nineteenth and twentieth century science, we view our lives and our world in a framework of competition, limited resources, and a struggle to survive.[2] No matter our religious or spiritual outlook, for the most part we take our world to be what we see and touch, what we observe and quantify. We embrace a mechanical interpretation of life, a materialistic vision of creation.

The world of tomorrow will also be a product of our worldview. Only, as our future unfolds, we will not see our world and ourselves in predominantly materialistic terms. It is my belief that we will embrace what in popular circles is called the *evolution of consciousness* vision of the universe. In this developing understanding of creation proposed by the scientist and philosopher *Pierre Teilhard de Chardin* in the early part of the twentieth century and refined and expanded on in the first book in

1. See *Book 1, Evolution of Consciousness,* chapter 1.
2. See *Book 2, Economics of Fulfillment,* chapters 1, 2, 3, and 4.

this series—*Threshold to Meaning: Book 1, Evolution of Consciousness*—evolution is more than what we may think. It is the process whereby the universe advances to greater autonomy and consciousness. It is how time progresses, how the universe matures. It is the way creation advances to more refined and inspiring states of existence. It is how the universe, and all that it has and will embody, perfects itself.

In the course of this perfection, evolution takes place through a clearly evident mechanism of change, the *creative process*. As such, we no longer see evolution as advancing gradually, or linearly—accounted for by the *big bang* model of cosmic formation, the *natural selection* model of organic change, and other external theories, though these theories may be useful to a point and have merit in limited situations—but through a process of trial and error, alternative evolutionary directions, and creative cycles and thresholds. We progress through periods of stability and cyclic conditions marked by a buildup and collapse of uncertainty that culminates in the crossing of a threshold to a higher plane of existence and to a new line of evolutionary advance.

As important, we realize that the creative process operates on two opposing but complementary levels. Like the writer who while drafting a story creates and then reshapes and discards text as he or she coaxes the story to its final form, the *leading arrow* of the creative process propels evolution forward, to states of greater autonomy, complexity, and consciousness. The *trailing arrow* reshapes and discards prior evolutionary forms in support of evolution's overall advance. The creative process builds on and creatively discards the old to create the new.[3]

Over the course of the universe's advance, the majority of evolutionary thresholds were small, barely discernible transformations that delineated the flow of time. Others were large—sweeping developments built on the cumulative steps of smaller ones and that thrust the universe to altogether new levels of evolutionary achievement. As presented in the first book, and as *Figure 1* shows, such thresholds allow us to divide evolution into major evolutionary periods separated by major evolutionary thresholds. We called these periods the universe's ages of *Emergence*, *Structure*, *Life*, *Understanding*, and *Fulfillment*.

---

3. See *Book 1, Evolution of Consciousness*, chapter 6.

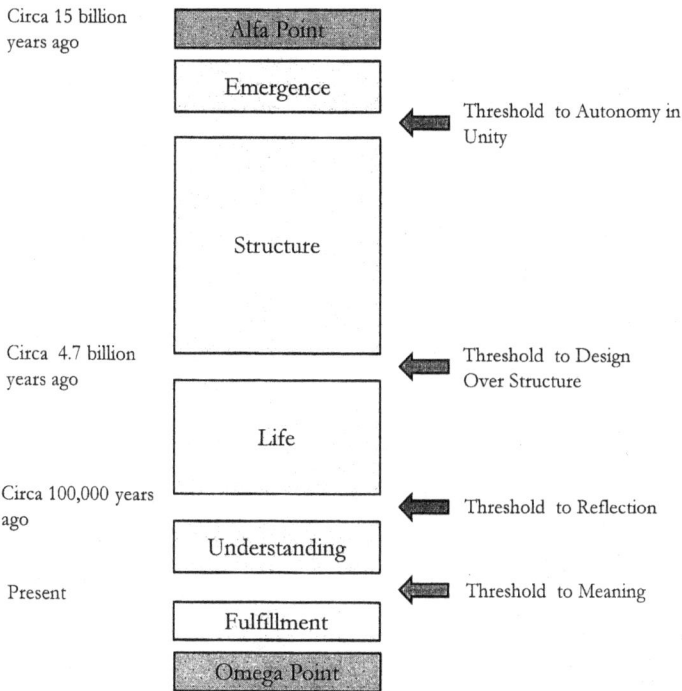

Fig. 1. Major Evolutionary Periods and Thresholds. From the universe's origin in emptiness, or its Alpha Point, to the universe's conclusion in a sustained state of fulfillment, or its Omega Point, we can divide time into five major stages of evolutionary advance, each separated by a major evolutionary threshold.

Circa one hundred thousand years ago, evolution marked the turning point to *Reflection*,[4] the moment when the universe through the evolution of the human psyche became conscious of its existence and entered its most recent evolutionary period, the age of understanding. Today, the universe faces another, no less significant turning point, the breakthrough moment that as I write these words thrusts evolution beyond the age of understanding into its final evolutionary period, the age of fulfillment. Today the universe crosses the threshold to meaning.

4. The date of one hundred thousand years as the time of the universe's threshold to reflection is tentative and may prove to be earlier. The advent of the human ability to think and learn reflectively is generally thought to coincide with the appearance of funeral rites in the archeological record. See *Book 1, Evolution of Consciousness*, chapter 10.

Like our transcendence to reflection, the crossing of this threshold unfolds within the human being. It represents our transformation to a higher state of existence, our leap to a new level of awareness. Endowed with meaning, we are not only aware of our consciousness we are aware of our evolution. In a manner no less intrinsic than our sense of self, we internalize the course that brought us into existence and the immediate and ultimate future to which this course leads. We embody the direction and purpose of our creation, the reason for our being.

This review of the evolution of consciousness perspective and of humankind's threshold to meaning allows us to reframe a central idea in a way that meets our present needs. The idea is that of *human motivation*. As reflective beings, we are driven to grow and learn. Even a cursory look at history and at the events that every day unfold in the world make it clear that our motivation is to move forward, to expand the reaches of thought, to push outward the limits of what it means to be human. We are driven to evolve, to reinvent ourselves, to create within ourselves a higher state of being, to advance to a state of greater insight and awareness. Moreover, as did the twentieth century psychologist Abraham Maslow who saw human motivation as a progression from basic needs to higher needs of "self-actualization," we can categorize the human desire to advance to greater levels of awareness, understanding, and creative expression. We can form a hierarchy of needs.

As Maslow understood, practical reality compels us to focus our energy on our material well-being. The first level in our hierarchy of human motivation is the most pragmatic. In today's global economy, our most pervasive needs are all too often associated with our physical embodiment. Social demands and practical realities focus our attention on our material well-being. They center our thoughts on *economic* matters.

In the absence of the material goods necessary for survival and a reasonable level of comfort and security, our economic motivation reduces to the simplest terms. When hungry, we seek food. When cold, we seek clothing. When caught in the elements, we seek shelter. When we or those whom we care for are malnourished, inadequately dressed, and exposed to the environment, we are compelled to satisfy the immediate wants of subsistence.

Yet, in today's economy, our material wants are not clearly defined. We cannot separate our need for the material necessities of life from the mechanics of the global economy. In technologically developed regions of the world, few of us grow or hunt our food, and few of us make our clothes and build our shelters. We are a cog in the economic works, a component in the vast interplay of connected variables that characterizes the function of our socialist-capitalist economic environment. Our survival depends on more than food, clothing, and shelter.

Our material needs are further complicated by the relationship that has evolved and that we choose to maintain between our economy and the social structure of our community. For many, the work we do establishes our position in society. It is not enough to have a job that allows us to buy the goods we need to survive. We must have a job that defines our social standing and that allows us to purchase the trappings that express our social standing. In different ways and to different degrees in every nation and part of the world, social structure is entwined with our economic values and practices, with our material hopes and dreams.

When we are fortunate enough to reach the point in our lives where we experience the fulfillment of our economic needs—and when we are fortunate enough to grasp the nature of these needs and by doing so to gain the perspective to look beyond—our desire to better ourselves and our world crosses the threshold to a higher level of expression. As we described in the earlier books in this series, the creative process must have room to work. To grow and learn, we must provide for ourselves and for those around us the flexibility to try, fail, and try again. When we experience the fulfillment of our economic needs, and acquire the character to look outside the box of materialistic desire, we direct our creative energy to creativity itself. Our goal becomes to optimize the environment of creativity. We are driven by a higher calling, by a drive that throughout history has propelled change. We are driven by the need to establish and thrust to ever higher levels of expression the human experience of *freedom*.

We can think of human progress, and specifically the human pursuit of freedom, as influenced by two opposing forces. The need to advance that we spoke about earlier in the chapter drives us forward in time, to dream new ideas, to invent new ways to do things, to create more nestled and intimate social arrangements. On the other hand, a need we all on occasion feel—the desire to maintain the status-quo—encourages us to look back in time. We are motivated to value the less evolved. We seek solace in the collective social ethics of the past. We all struggle to accept new ideas and new ways to do things, and prudence in doing so is an admirable quality. Yet, we at times reach a point where our attachment to old ideas and old ways hinders our development. We arrive at this state when we direct our inventiveness and creative energy to justify ideas and ways for no other reason than they may once have been worthwhile. We call the process whereby we direct our natural ambition and creative power to maintaining the old beyond its usefulness *stagnation*.[5]

Within the individual, stagnation reveals itself as insecurity and as the ego that results when we too tightly associate our sense of self with beliefs kept alive beyond their time. With no firm convictions and nothing new or worthwhile to say, a politician will attack his opponent. Within humanity, stagnation reveals itself as dogma, racism, ethnocentrism, and fundamentalism. Is not the turmoil that today grips the Islamic world a battle between modern and ancient religious interpretations? Many historians and philosophers have reduced the totality of the human experience to the quest for freedom, and many have reduced the quest for freedom to a war between the past and the future. Many have interpreted the ascent of civilization as a struggle between the novel and the stagnant, the creative and the status-quo, the road to tomorrow and the quagmire of the present. At least to this point in time, does not human progress and social transformation come down to a struggle between the values of freedom, creativity, and individuality and the values of conformity, government, and central control?

In our struggle for freedom, we face many obstacles. Legal and political barriers limit what we can say and do. Social norms restrict our behavior in ways that are at times appropriate and at times have little relevance. As we described in *Threshold to Meaning: Book 2, Economics of Fulfillment*, the greatest obstacle to our freedom, however, is economic.[6] No matter what economic ideology we embrace—socialism, capitalism, or the blend of these practices every nation employs—to a remarkable degree the human community is driven by ideals of power and material acquisition. In this economic framework, the freedom of the individual may by limited directly: by the will of leaders imposed through a nation's legal system and the police power of the state. The freedom of the individual may also be limited by economic concerns. In socialism-capitalism, freedom equates with opportunity and opportunity is a function of capital

---

5. See *Book 1, Evolution of Consciousness*, chapter 11.

6. See *Book 2, Economics of Fulfillment*, chapter 1.

and the individual's access to capital. Capital provides the means to direct our creative energy as we see fit: to raise a family, to earn an education, to start a business, to seek a fulfilling job, to bring into production an invention. Factions compete for capital and the power it represents and by doing so limit the freedom of the individual.[7]

Yet, today, the human community has taken the first step beyond its preoccupation with economics. We have opened ourselves to the possibility of an existence free from economic limitations, of a way of life unbounded by economic constraints. Central to this transformation is *economics of fulfillment*. The objective of the economics of fulfillment ideology is to create the social and economic conditions that allow every individual to evolve to his or her highest state of being and that by doing so allow humanity to evolve to its highest state of being. Economics of fulfillment is a system of hyper-individuality, of hyper-opportunity, of hyper-free-enterprise. It is a philosophical path to greater fulfillment by way of greater personal freedom and the acceptance of the responsibility this implies.

As profound as our quest for freedom may be, as forceful as it may have been in shaping our history and politics, as essential as it will be to our future, our need for freedom is a step along the way in our hierarchy of needs. When we fulfill our need for freedom, we unleash our creative drive to manifest in a yet more evolved way. Our need to reshape our world and reinvent ourselves reveals itself in its purest form—as the drive for *perfection*.

At the heart of our quest for perfection is the need to make sense of our world. By way of our ability to reflect, we maintain an evolving vision of ourselves and our universe. The human yearning for perfection embraces our desire to understand. It incorporates our urge to grasp the universe's nature and purpose. It is the embodiment of our evolution to greater autonomy and consciousness. The universe's drive for perfection unfolds within us.

The universe's drive for perfection also unfolds outside of us. Inseparable from our need to refine our vision of the world is our need to reshape our surroundings. Motivated by the calling of perfection, we no longer see the outcome of our actions in strictly economic terms. Our goal is not to get the job done as fast and cheaply as possible and to maxi-

---

7. Ibid., chapter 4.

mize our profit or, to be more accurate, the profit of our employers. It is to better the world. Whatever our task, perfection is our choice to perform it to the best of our abilities. A nurse strives to interact with her patients in the most satisfying manner. A teacher strives to nurture her students in the most stimulating way. A homebuilder strives to build the perfect house and to make each house better than his last. Perfection is our road to fulfillment. At times, we all strive to do our best. We take satisfaction in reshaping the world within our reach.

Our drive for perfection also manifests in our desire to interact with one another in a more meaningful way. Driven by the need for perfection, family takes precedence over economics and personal gain, and community takes precedence over politics. As the individual evolves to great individuality and substance of character, we gain the ability to form stronger, more intimate social bonds. Correspondingly, society evolves from more to less collective forms.[8] Perfection is the purest expression of our evolution. We seek perfection by realigning and strengthening our connections to one another in ever more nestled, intricate, and fulfilling arrangements.

We strive for perfection within ourselves. We strive for perfection through our activities and through our social interactions. But perfection does not mean to evolve without effort. Driven by the need for perfection, we direct our creative power in a way we find satisfying—with all the trial, error, and uncertainty inherent in the creative process. Perfection is our desire to reshape ourselves and our world into ever more evolved forms, redefining the ideals to which we aspire along the way. Empowered by freedom, we are perfectionists, but we do not lose ourselves in the obsession of perfection. Our quest for perfection is our evolutionary path, our road to betterment. The individual seeks self-perfection. Humanity seeks perfection of life on earth.

Human motivation displays a hierarchy of needs. When our economic needs are fulfilled, we strive to create freedom. Empowered by freedom we strive to create perfection. This hierarchy is not absolute or clearly defined. At one moment we may devote our energies to economic matters, at another to freedom, and at still another to perfecting our surroundings. Our hierarchy of needs is not rigid or immutable, but it is a hierarchy none-the-less.

8. See *Book 1, Evolution of Consciousness*, chapters 5, 9, 10, and 11. See *Book 2, Economics of Fulfillment*, chapters 2 and 12.

Perfection

The world as we experience it is a product of our concepts and understanding. Today, humanity marks the turning point to a new level of existence. We internalize the universe's origin, nature, and purpose and transcend the divide to a consciousness of our evolution, to an internalization of all time past. We cross the threshold to meaning. Empowered by meaning, we know where we have been, where we are at, and where we want to be. We embody the wisdom to direct our creative energies beyond economic subsistence to freedom and, liberated from the demands of false limitations, to the perfection of ourselves and our world. Our task is to reengineer the earth's *urban and ecological infrastructure*. Our challenge is to draft the blueprint for reconstruction.

PART ONE

The Landscape

2

Pre-Reflective Landscape

A S ANY ENGINEER OR architect would do at the beginning of a project, in our task to draft a blueprint for reconstruction of the earth's urban and ecological infrastructure, we begin by studying the site on which we will build—in our case, the earth. In the book's first section, we look at how our planet and its biosphere have changed over time, incorporating into our discussion the impact of human activity on this evolution. This establishes the background to understand the economic and political forces at work on today's urban and ecological landscape, forces that we must be aware of to engineer tomorrow's urban and ecological landscape. In particular, it carries us beyond contemporary economic and environmental doctrine to a new approach to urban and ecological management—to the strategy-of-progress that, to create our design for tomorrow, the future will compel us to embrace.

As for the earth prior to humankind's crossing of the threshold to reflection one hundred thousand years ago, or what I refer to as its "pre-reflective landscape," we begin our overview with a geological description of our planet and expand our discussion to include life and its evolution, culminating with humankind and its evolution. This leads us to an observation that may challenge our existing view of nature. It is also an observation that is well-justified and central to our understanding of the future and to our new strategy of urban and ecological management.

Our planet is the third from the sun and the fifth largest in the solar system. Contrary to popular belief, the earth's orbit is not a true circle. Our planet travels in a slightly elliptical orbit with a mean distance to the sun of 149.5 million kilometers, or 92.9 million miles. The earth is also not a perfect sphere but slightly pear-shaped, with a 21 kilometer, or 13 mile, bulge at the equator, a 10 meter, or 33 foot, bulge at the north pole, and a 31 meter, or 102 foot, depression at the south pole.

Current theory suggests that the earth formed from a cloud of stellar gas and heavy element debris that drew in on itself to create the sun and planets. Radiometric dating places this event at between 4.6 and 5 billion years ago, near the end of what in *Threshold to Meaning: Book 1, Evolution of Consciousness* we defined as the universe's evolutionary period of structure.[1] At first, the earth was a relatively cool and homogeneous mass. Gravitational contraction and the radioactive decay of uranium and other elements, however, caused the earth to heat and melt. Heavier elements, primarily iron and nickel, sank to the planet's center. Lighter elements, primarily silicates, floated to the surface. By about 4.4 billion years ago, an atmosphere of methane, ammonia, nitrogen, hydrogen, water vapor, and carbon dioxide, with little gaseous oxygen, had accumulated. By about 4.1 billion years ago, water vapor in the atmosphere had condensed to blanket the planet in the first oceans.

As a result of the earth's formative processes, our planet is made up of five spherical layers: the *atmosphere*, the *hydrosphere*, the *lithosphere*, the *mantle*, and the *core*.

At present, the atmosphere is about 1,100 kilometers, or 700 miles, thick, with half of its mass in the lower 5.6 kilometers, or 3.5 miles. It consists of 78 percent nitrogen, 21 percent oxygen, 0.9 percent argon, and 0.03 percent carbon dioxide, with trace amounts of ozone, methane, hydrogen, and carbon monoxide. Depending on elevation and temperature, the atmosphere also contains varying amounts of water vapor—which range from saturation to, during the height of winter, only small amounts at the poles.

The hydrosphere is the layer of liquid water that covers most of the earth's surface. It consists of the seas and oceans, all lakes, rivers, inland seas, and underground waterways.

The lithosphere is a layer of hard, cold rock that extends from the earth's surface to a depth of about 100 kilometers, or 60 miles. It is made up of the *sialic*, or upper crust, which includes the continents, and the *simatic*, or lower crust, which forms the ocean floors and basins. The upper crust consists of sedimentary and igneous rocks, the latter typified by granite. The lower crust consists of darker, heavier igneous rocks, in particular gabbro and basalt. The lithosphere also includes the upper mantle, which is composed of iron and magnesium silicates and is separated from

---

1. See *Book, Evolution of Consciousness*, chapter 7.

the crust by a seismic discontinuity called the *Moho*. In terms of elemental composition, oxygen is the most abundant element in the lithosphere, followed by silicon, aluminum, iron, calcium, sodium, potassium, and magnesium.

The mantle as a whole extends from the base of the crust to a depth of about 2,900 kilometers, or 1,800 miles, and, similar in composition to its upper shell, is thought to be made-up largely of oxides of iron, silicon, and magnesium.

Enclosed by the mantle is the core. The core has an outer shell that extends down about 2,225 kilometers, or 1,380 miles, and is thought to be made up of iron with a small percentage of nickel and other elements and to have peaks and ridges on its surface that form where warm material rises. The core's inner shell has a radius of about 1,275 kilometers, or 795 miles, and is also thought to be composed primarily of iron. Geologists generally consider the core to be liquid but, due to the tremendous pressure that would be exerted at such depths, also theorize it to be solid near its center. The temperature at the center of the core is hot, estimated to be as high as 6,650°C, or 12,000°F.

One of the earth's most remarkable characteristics is the phenomenon of continental drift. The lithosphere, the earth's outermost solid layer, is broken into twelve major tectonic plates and is separated from the lower mantle by a zone of partially molten rocks called the *asthenosphere*. The intense heat in the core, thought in part to be the energy released by the decay of uranium and other radioactive elements, generates convection currents in the lower mantle. These currents apply shear forces to the asthenosphere, which transfers these forces to the tectonic plates. The movement of these plates causes the continents to travel across the earth's surface and the oceans to open and close. Convection currents also supply a flow of molten rock to the earth's volcanoes and mid-ocean ridges.

In addition to its overall structure, the earth displays a more subtle design, illustrated by its magnetic field. As anyone who has used a compass knows, however, the poles of the earth's magnetic field do not coincide with the earth's geographical poles. At present, the north magnetic pole is off the coast of Bathurst Island in the Nunavut region of the Canadian Northwest Territories. The south magnetic pole is near the Adélie Coast of Antarctica. The position of the earth's magnetic poles shows daily and annual variations and a periodic variation every 960 years. On occasion, the earth's magnetic field also reverses polarity. Studies of residual magnetism in rocks and of magnetic anomalies on the ocean floors show that the polarity of the earth's magnetic field reverses frequently in geological time—every few hundred thousand years and as many as 170 times in the last 100 million years.

In addition to a magnetic field, the earth maintains three electrical systems: one in the atmosphere, one that flows parallel to the earth's surface, and one that transfers an electric charge between the atmosphere and the earth's surface.

In the atmosphere, electricity results from ionization caused by solar radiation and by the movement of ions on the atmospheric tides. The ionization, and consequently the conductivity, of the atmosphere increases with altitude. Between 40 and 400 kilometers, or 25 to 250 miles, ionization forms a conductive spherical shell called the *ionosphere*. The ionosphere reflects certain radio signals back to the earth and absorbs certain spectrums of electromagnetic radiation from the sun.

On the earth's surface, physicists have identified a system of eight current loops located on either side of the equator and a series of smaller loops located near the poles. These currents are thought to originate in the earth's core, which acts like the armature of a huge electrical generator. Thermal convection currents in the core move molten metal in loop patterns relative to the earth's magnetic field. This movement produces electrical currents that mirror the earth's thermal convection currents.

Like any current, the transfer of electricity between the earth and the atmosphere results from a voltage differential. The surface of the earth has an overall negative electrical charge. The prevailing view is that this surplus of electrons attracts positive ions from the atmosphere during fair weather. The downward flow of positive ions is then balanced by a return flow to the atmosphere during thunderstorms.

As brief as our description of the earth may be, it makes clear one important point. Our planet is not a static entity. Volcanism, continental drift, magnetic fluctuations, and electrical currents dramatize our planet's vibrant nature. Our earth demonstrates the characteristics of the creative process—behavior that is cyclic and dynamic.

This becomes even more apparent when we expand our look at the earth to include the biosphere, or its layer of life. By about four billion years ago, the first single-celled life forms had emerged. At the time of this development—at the moment evolution crossed what we referred to in *Book 1, Evolution of Consciousness*, as the turning point of *design over structure*—the cosmos and all that came before fell into the realm of evolution's trailing arrow. Evolution's leading arrow locked in the design of life.[2]

In life's ascendance, evolution advanced on three distinct levels: First, the creative process progressed through the design of the life form itself. Unicellular life gave way to multicellular life. Multicellular life evolved into species that embodied circulatory systems, reproductive systems, and nervous systems. From the standpoint of consciousness, life forms displayed a growing awareness of their surroundings and a greater flexibility in their behavior. Life became more free and animated, more lifelike.

Second, the universe perfected itself through the design of social structure. To some degree even the most primitive life forms maintain relationships between members of a species. The more advanced, and thus the more consciousness, the species, the more intricate the relationships.

In ways we have only begun to understand, bacteria and other primitive life forms communicate through chemical secretions to coordinate reproduction and their relationship to one another. Further up the evolutionary ladder, the marine polyps whose calcareous secretions form coral display a still more complex social organization.

A few rungs higher, the ant colony displays a mechanistic pattern of social interaction. In a typical colony, infertile worker ants gather food and build and defend the nest, and male ants impregnate the queen whose sole duty is to produce offspring. The colony may be large and demonstrate a division of labor, but the bonds between the members are weak. Social structure is uniform and mechanical.

As we continue up the evolutionary ladder, the uniformity and mechanistic quality of social organization decreases. This is particularly apparent on the level of the mammal. In the bison herd, females with young form tight bands and males form their own groups. In the lion pride, adult males, females, and young maintain closely-knit subpride relationships within the highly organized pride structure. The wolf pack displays an even more complex organization comprised of parents and offspring. When we reach the level of the ape and monkey, species such as the gibbon display a rudimentary family and community structure.

2. Ibid., chapters 7, 8, 9, and 10.

The more conscious and thus the more evolved and autonomous the organism, the stronger the social bond it can form. As more advanced life emerged, the social structure it displayed became more nestled, interwoven, and strongly bonded, less *collective*.[3]

Third, the universe perfected itself through the design of the biosphere—through the refinement of interactions between species to form ecosystems and between ecosystems to form the larger whole that defines life on earth. A species of single-celled entities that consumes oxygen and releases carbon dioxide interacts with a species that consumes carbon dioxide and releases oxygen. A predator interacts with its prey, and its prey interacts with species on successively lower levels of the food chain. The universe advanced through the complexification of life's connections.

Over the course of organic evolution, the forward arrow of the creative process jumped between levels. At one moment, it was locked in the advancement of the organism, at another in the refinement of social structure, at another in the perfection of the biosphere—in the organization of ecosystem relationships.[4] In addition, a change on one level would lead to changes on other levels. An increase in organism consciousness, for example, would result in a decollectivization of the organism's social structure and in a change in the way the species interacted with other species.

By about 3.4 billion years ago, the first cells capable of conducting photosynthesis using a lower, less energetic wavelength of light than ultraviolet had emerged. This made it possible for cells to create carbohydrates from carbon dioxide and water while giving off oxygen and led to a gradual buildup of oxygen in the atmosphere and to the formation of an ozone layer.

By 1.4 billion years ago, atmospheric oxygen had reached the level needed for cells to maintain an aerobic, or oxygen based, metabolism. The use of oxygen increased the cell's ability to downgrade energy stored in the glucose molecule, and metabolism became more efficient. This opened the way for two developments that would forever change the face of life. These were sexual reproduction and the internalization of death,

---

3. The basis for the idea of collectivity, and for the trend of evolution from more to less collective forms of social organization, can be traced as far back as the formative stages of the universe's cosmic evolution. See *Book 1, Evolution of Consciousness*, chapter 5.

4. See *Book 1, Evolution of Consciousness*, chapters 8 and 9.

or the emergence of lifespan. In the evolution of consciousness view, sex functions as a mechanism to increase the rate of the universe's trial and error. In contrast, death and lifespan function on evolution's trailing arrow, as a mechanism to discard life's obsolete designs.[5]

By 800 million years ago, social structure had progressed to the cell colony. *Heterotrophy*, or the feeding by one species of life on another, had become widespread, and the organization of the biosphere reflected the complexity of this interaction. By 750 million years ago, the creative process had exhausted its potential to advance through the evolution of unicellular life, and the universe crossed the threshold to the multicellular life form.

At the onset of the *Paleozoic* era, 540 million years ago, the earth's surface consisted of a vast ocean from which rose various landmasses. Most of these landmasses, or protocontinents, were situated in the tropics and southern hemisphere. Life was confined to the seas, which teamed with algae, seaweed, worms, sponges, mollusks, and other invertebrate organisms.

By about 500 million years ago, the protocontinents that would one day form Europe and North America had come together to create a vast continental area, much of which was submerged beneath a shallow sea. In this and other aquatic regions, corals and clams flourished, as did primitive armored fish and other vertebrates. By about 450 million years ago, plants had colonized land, followed, fifty million years later, by animals. Not long after, the first vascular, or nutrient circulating, land plant had developed, as had the first air-breathing animal. At the same time, rays, sharks, and other cartilaginous fishes roamed the world's oceans. By about 370 million years ago, flying insects had emerged and forests of ferns and woody plants that grew to tree-like heights covered much of the land. Ultimately these forests would decay to create many of the planet's oil and coal reserves.

By about 250 million years ago, all of the earth's major landmasses had come together to form a single super-continent called *Pangaea*. Not long after, many fishes and a large number of other species died out in what some paleontologists consider the greatest mass extinction of all

5. Ibid.

time. This extinction marked a fundamental shift in the developmental thrust of the biosphere's evolution.[6]

As we have said, biologists, paleontologists, and others in the scientific community embrace a largely external, mechanistic interpretation of life and its evolution. The prevailing view is that organic evolution took place through random forces: through Darwinian natural selection. Despite fossil evidence that documents the pattern of cycles and thresholds we would expect to see from the creative process,[7] the evolution of consciousness view and the mechanism of the creative process have only begun to enter scientific consciousness. Unaware of the creative process and therefore unable to account for events such as the mass extinction at the end of the Paleozoic as an intrinsic aspect of evolution, scientists look for an explanation wherever they can find one.

Traditionally, the explanation of choice has been asteroids, though in recent years volcanoes and geological activity have gained ground. Other theories proposed include diseases and super-hurricanes. In the asteroid theory, the mass extinction at the end of the Paleozoic was the result of the earth's collision with a celestial object that released sufficient energy to cause global climatic upheaval. The evidence to suggest that an asteroid hit the earth at the end of the Paleozoic is controversial. The evidence to suggest that an asteroid struck with sufficient force to cause climate change and bring about mass extinction is even weaker. But there is evidence to suggest that asteroids have hit the earth. When we consider the number of asteroids observed in the solar system and the age of the earth, it seems reasonable that our planet has been struck on numerous occa-

---

6. Though the fossil record is open to interpretation, paleontologists widely consider the mass extinction at the end of the Paleozoic to be the most extensive. They also widely consider the number of species and ecosystems, and thus the diversity and complexity of the biosphere, to have been at its maximum just prior to that event. In the Mesozoic, there were a number of periods when biosphere complexity increased and a number of mass extinctions, including the one that took place sixty-five million years ago and brought about the end of the dinosaurs. None of these periods of rise and fall, however, is generally thought to have exceeded the Paleozoic event. From an evolution of consciousness standpoint, the Paleozoic extinction marked the point when the forward thrust of the universe's overall creative process shifted from a focus on the evolution of life in its entirety—the biosphere—to a focus on the evolution of the organism and organism social structure. At some point in life's advance, this shift in evolutionary thrust clearly took place. The mass extinction at the end of the Paleozoic is the logical marker.

7. See *Book 1, Evolution of Consciousness,* chapter 9.

sions, especially during its formative years.[8] As for the geological theory of cause, in this view the mass extinction at the end of the Paleozoic was the result of catastrophic volcanic eruptions that thrust vast amounts of sulfur and other toxic substances into the atmosphere. The evidence to support this theory is better, but as we described in *Book 1, Evolution of Consciousness,*[9] a random event such as an asteroid collision, a chain of massive volcanic eruptions, or, even more speculative, disease outbreaks and massive hurricanes was not the driving force behind the Paleozoic extinction or behind any other significant evolutionary development.

The Paleozoic extinction took place because the universe had exhausted its evolutionary direction. The Paleozoic extinction marked a turning point, an evolutionary crossroads. During the Paleozoic, the biosphere had attained such a degree of complexity that, arguably, it represented the most sophisticated level of organization it would ever achieve. At the end of the Paleozoic, the leading edge of the universe's creative process drew away from the design of the biosphere. To an extent not before seen, it locked in the further development of the organism and in organism social structure. The ecological diversity and complexity of the Paleozoic was no longer needed.

The mass extinction at the end of Paleozoic ushered in a new era, the *Mesozoic,* or the age of the reptile. During the Mesozoic, which began 245 million years ago, Pangaea split into two continents, the southern called *Gondwanaland,* the northern called *Laurasia.* Gondwanaland then split into what would become India, Africa, Australia, Antarctica, and South America, and Laurasia split into what would become Asia, Europe, and North America. Biologically, the Mesozoic saw the rise of many life forms, but none that has captured our attention more than the dinosaur.

The earliest dinosaurs were comparatively small creatures that ran on their hind feet and balanced their bodies against the weight of enormous tails. By 195 million years ago, the various species we are most familiar with had begun to emerge. These included the armor-plated stegosaurus, the rhino-like triceratops, the massive two-footed carnivore tyrannosaurus rex, the long-necked vegetarian apatosaurus, and the great winged reptile pterodactyl.

8. Asteroids may have played an important role in the formation of the earth's early oceans and atmosphere.

9. See note 7.

At the end of the Mesozoic era, 65 million years ago, another mass extinction swept the planet. Among the animals to vanish was the dinosaur. Locked in evolution's trailing arrow, such a creature was no longer necessary for life's advance. The earth had entered its most recent geological era, the *Cenozoic*.

The Cenozoic opened on a warm world with tropical forests that spread further north and south from the equator than they do today. It was an era defined by the rise of the mammal and by the mammalian order to which we belong—the primates. The creative process had locked on the evolutionary line that would lead to life's highest organic expression—the human biological form.

At the beginning of the Cenozoic, several major groups dominated the mammalian class. The various species that made-up these groups consisted of relatively small creatures, none that exceeded the size of the modern bear. All were four-footed, and most had mussels, slim heads, and five toes on each foot.

By about 54 million years ago, mammalian life had seen significant changes. The evolutionary ancestors of the horse, camel, rodent, and rhinoceros had emerged, as had the first aquatic mammals, ancestors of the modern whale. By about 38 million years ago, true carnivores had appeared—animals that resembled modern dogs and cats. By about 26 million years ago, various grazing species had evolved, as had their predators.

The mammalian class also included the primates. Curiously, the first primates were small, rodentlike creatures that resembled modern moles and shrews. But like all primates, they had refined vision with good depth perception and a comparatively large brain with a fissure between their first and second visual areas. Around 50 million years ago, the primates branched into two suborders, traditionally classified as *prosimians* and *anthropoids*. The prosimians changed little over time and became the modern loris, lemur, and tarsier. The anthropoids had a more colorful evolution. By 38 million years ago, the anthropoid line had branched. One shoot gave rise to the new-world monkey. The other gave rise to the old-world monkey, gibbon, and orangutan and to a number of early apelike species.

Between 7 and 10 million years ago, this line also branched. The shoot that would lead to the human line broke away from the shoot that would lead to the gorilla and chimpanzee; and, by about five million years ago, species clearly human in form had emerged. Classified in the genus *Australopithecus*, as opposed to the genus Homo to which we belong, these creatures walked upright and had small brains and protruding brow ridges.

By about two million years ago, a new creature roamed Africa—one too primitive to be called human, too advanced to be called Australopithecine. Anthropologists call this creature *Homo habilis*. Homo habilis looked much like the more evolved of the Australopithecus species but had slightly more refined facial characteristics. It also had a larger cranial capacity and made the first stone tools.

By about 1.7 million years ago, Homo habilis had disappeared, replaced by a creature still closer to ourselves, *Homo erectus*. Anatomically, Homo erectus was taller than Homo habilis though still shorter than a modern human being. It possessed a larger brain, a higher forehead, smaller teeth and jaw, and a less distinct ridge above the eyes. Homo erectus ranged from Asia to Africa to Europe and made a variety of stone awls, anvils, burins, scrapers, choppers, and hand-axes. Homo erectus lived in caves and made huts not unlike those constructed by modern hunting and gathering peoples such as the Khoi-san, or African Bushman. Homo erectus also hunted big game and killed animals as intimidating as the now extinct cave bear—larger than the modern Kodiak. The species also cooked its food, for it had learned to harness fire.

After more than a million years of anatomical refinement, the decedents of Homo erectus, archaic forms of the modern human being—of which the species Homo Sapiens Neanderthalensis, or Neanderthal, is the best known—roamed the earth. Of the evolutionary advances that took place in our archaic forms, one is paramount. Our ancestors achieved the nervous system complexity to support a new form of thought and learning. The anatomical structure—or structure of perception as we referred to it in *Book 1, Evolution of Consciousness*—was in place to support a profound leap in consciousness. At an evolutionary turning point that took place about one hundred thousand years ago,[10] the universe marked the development we have spoken about before. It crossed the threshold to reflection, and humanity set out to understand itself and its universe.[11]

10. See *Book 1, Evolution of Consciousness*, chapters 10 and 11.
11. Ibid.

This turning point also marked a new era in the evolution of the biosphere and brings us to a conclusion about the present direction of life's evolution that is central in our pursuit of urban and ecological reconstruction.

Conceivably, the biosphere reached the height of its diversity and ecological complexity just prior to the Paleozoic extinction, 245 million years ago. Based on an observed simplification of ecosystems and ecosystem interactions visible in the fossil record, and backed by the evolution of consciousness view and the dynamics of the creative process, we can argue that biosphere evolution subsequent to that event took place largely in support of advances in organism anatomy and social structure. There were periods when biological diversity vastly increased, and there were a number of mass extinctions, including the sixty-five million year ago episode defined by the extinction of the dinosaur. As a whole, however, the direction of the biosphere's evolution was toward decline.[12]

The threshold to reflection thrust this pattern of trailing edge creative activity to a new level. As did the cosmos before it, life itself fell into the realm of evolution's trailing arrow. With the threshold to reflection, the evolution of the biosphere no longer took place in support of the development of more advanced organic forms. The evolution of the biosphere took place in support of what had become the leading edge of the universe's advance—the human drive to understand itself and its world. On evolution's trailing arrow, the direction of the biosphere's evolution solidified. As of approximately one hundred thousand years ago, the evolution of the biosphere was, as during no period before, defined by decomplexification, by a reduction in the number of species and ecosystems.[13] This is not to say that new species did not emerge and in areas biological diversity did not increase. When we look at the biosphere in its totality, however, the direction of evolution subsequent to the dawn of reflection has been toward a decline in complexity. The biosphere has never been in a sustained state of equilibrium. All of humankind's reflective experience—all of science, all of philosophy—has taken place in an era of biosphere simplification.

12. Ibid., chapters 11 and 12.
13. Ibid., chapters 9, 10, 11.

3

Reflective Landscape

WITH THE THRESHOLD TO reflection one hundred thousand years ago, the direction of evolution on the level of the biosphere entered an era of simplification: an era defined by the falling away of the species and ecosystems not needed to support what had become the universe's forward thrust in evolution—humankind's advance to greater knowledge and understanding.[1] As we said in the previous chapter, this does not mean that new life did not emerge. Bacteria change form to accommodate new conditions. Through domestication, we create new strains of plants and animals. Little more than a generation after Mount Saint Helens erupted in 1980, new forests have taken root on land that was devastated ash fields. Subsequent to the universe's threshold to reflection and the climax of the evolutionary period of life, however, the thrust of the biosphere's evolution has been toward decomplexification.

During the vast majority of this period, human interaction with what I call the earth's "reflective landscape" was limited. We existed as nomadic hunting and gathering cultures and interacted with our environment much as any other species in an ecosystem. This relationship changed about twelve thousand years ago when we began to leave our nomadic, hunting and gathering way of life and build the first permanent settlements.[2] Humanity's relationship with its environment further changed when, after this development, we initiated the practice of agriculture.

1. The reader not familiar with the earlier books may find the idea of the biosphere reshaping in support of humankind's advance to greater knowledge and understanding disconcerting. This is a central conclusion reached in *Book 1, Evolution of Consciousness*, where it is developed and justified at length.

2. The date of the first permanent settlements is generally thought to be about twelve thousand years ago. Though controversial, some anthropologists place this date as early as seventeen thousand years ago. In that the earliest settlements where primitive, there is also uncertainty as to what constitutes a permanent place of occupation as opposed to

As we established in the earlier books, urbanization was an outcome of the universe's evolution to greater consciousness, expressed as the human drive to create less collective and more intimate forms of social organization. Early hunting and gathering cultures maintained a communal way of life where, with the general exception of group leaders or some form of a ruling elite, members had roughly equal standing in society. Over time, groups organized to form tribes, and tribes organized to form allegiances of tribes. Driven by greater individual autonomy and consciousness, we interacted in more closely associated ways and created more complex and interwoven societies. As a result, we found it necessary to live in the close proximity provided by a geographically anchored community.

Moreover, as settlements grew in size, natural ecological systems in the vicinity could no longer provide for our material needs. As a consequence, we began to modify these systems. We adopted *agriculture*, or the practice where through the input of labor and organic material we simplify species relationships to form an artificial ecosystem, or one where species interact in a controlled way and as a result produce a desired product. As we described in the earlier books, agriculture was not the invention, or "basic adaptation," that made urbanization possible. To cite a classic example, excavations in Iraq tell us that the city of Shanidar was established about eleven thousand years ago, a thousand years before the first evidence of sustained food production in the region. Even our archaic ancestors, Homo sapiens Neanderthalensis and others, probably understood the concept of plant germination, but they had no reason to apply it. Why grow food when you can pick it up in your wanderings? Agriculture did not lead to urbanization. Agriculture made it possible for us to sustain urban growth and our drive to create less collective social orders.

With urbanization and agriculture, natural ecological systems existed alongside and interacted with artificial ecological systems. The biosphere's devolution continued; and, for the first time, human activity played a discernible role. To a degree, our conscious actions influenced the direction of evolution's trailing arrow. In part, we were responsible for molding what we can now properly call the earth's urban and ecological infrastructure.

---

a village or other site lived in for part of the year to follow the migration of game or for some religious or other purpose.

This reshaping of our world was not limited to the countryside but also took place in the community. The first settlements were little more than clusters of huts or other shelters built near a stable source of food or water or in an area with trade or religious significance. In humankind's ascent out of collectivity, the earliest urban communities grew and aligned to form larger communities. As time passed, the family differentiated to become the core social unit, which formed the basis for less collective forms of the community, which, as time passed, grew and aligned to create the first cities. Cities formed allegiances to create states. States came together to form nations, and nations came together to form empires. As settlements grew in size and influence, their design become increasingly important.

In pre-dynastic Egypt, the Nile plateau saw a gradual transition to urban life. Communities were small, linked by the Nile itself. Life centered on the farmland and on the system of irrigation canals that sustained it. As urbanization expanded, conflict became common as alliances broke and formed in our drive to create less collective forms of social structure. In this era of strife, cities were compact, enclosed by massive defensive battlements. In the Old Kingdom, and later in the Middle Kingdom and in the New Kingdom, social conditions were at times more stable. Cities became larger and more assessable to outsiders. Builders worked with granite and limestone and constructed monumental works of civic and religious architecture.

In Ancient Greece, the architect Hippodamus of Miletus, who planned the settlements of Priene and Piraeus, emphasized a geometric layout. Streets were arranged in a grid, and citadels were oriented to create a sense of balance. Housing was integrated with cultural, religious, commercial, and defense facilities. In Greek city planning, beauty was as much of a consideration as functionality. Cities were designed to be flowing and inspiring, with crowning structures built to honor the gods and to elevate the spirit. Many cities were constructed in mountainous regions, where their layout was integrated into the natural landscape.

Roman society was driven by ideals of power and grandeur—by the need to dominate man and nature. The layout of buildings emphasized symmetry, function, and a growing understanding of geometry and engineering. Colonial cities were planned as military camps, with buildings laid out in a grid surrounded by square or rectangular defensive walls. In the eyes of its rulers and citizens, Rome was more than an empire, and the city of Rome was more than the heart of a vast expanse of ruled territory. Rome was the center of the universe—an idea, a vision. In the city of Rome, streets were grand, paved with stone and brick. Structures were ornate, massive, imposing.

The evolution of our urban environment is also evident in the architecture and engineering incorporated into the structures we built. In pre-industrial cultures, construction techniques were dominated by the materials at hand. In forested areas, wood was an important building material. In areas where wood was scarce, builders tamped mixtures of clay and soil into molds to make bricks. The bricks were sun dried or, as the craft developed, fired in kilns, which increased their longevity and water resistance. Where available, no material was more desired than stone. Cut and shaped with stone bores, wedges, and hammers or with copper, bronze, and iron implements, marble and limestone were beloved for their workability, granite for its hardness.

The first monumental works of religious architecture done in stone rose along the Nile River about 4,700 years ago. Most notable was the Step Pyramid of King Zoser at Saqqara. Built in Egypt's third dynasty by *Imhotep*, the world's first documented architect, the Step Pyramid is one of the oldest surviving stone structures. As its name implies, the pyramid rises in distinct steps that reflect a construction technique where lower levels of the pyramid were used as the staging area to build higher levels. Later pyramids such as those at Giza show a more refined design, an evolution of construction skill and engineering knowledge.

Authors, mystics, and others have gone to great lengths to make us think that the construction of pyramids, in particular the building of those at Giza, is a mystery, in some way supernatural. In reality, the techniques used to build the pyramids are well established and supported by solid archeological evidence.

The Great Pyramid at Giza is 147 meters, or 481 feet, high and has a square base 230 meters, or 756 feet, on each side. To lay out a nearly perfect square, the builder needs nothing more than a grasp of basic geometry. As *Figure 2* shows, a foundation is square when four equal lengths of rope strung along the sides are positioned and adjusted in such a way that two ropes stretched from the corners and that crisscross in the middle are the same length—about one and two-fifths that of a side rope. So functional is this technique, it is used by builders to this day.

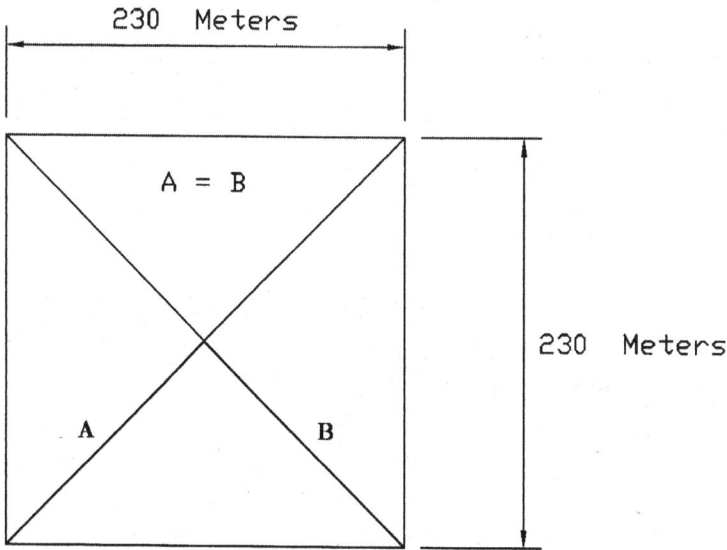

Fig. 2. Pyramid Layout. The foundation of a pyramid is square when four equal lengths of rope strung along the sides are positioned and adjusted in such a way that two ropes stretched from the corners and that crisscross in the middle are the same length—about one and two-fifths that of a side rope.

Excavation of a pyramid site is also a straightforward process. To level the site, simple devices that use water or gravity as a leveling agent— in principle the same as the surveying equipment used today—would have been sufficient to adjust the height of the key corner blocks. With these blocks leveled and positioned, workers could then drag the first tier of stone blocks into place and align them with the top faces of the cornerstones.

With a square and level foundation, subsequent tiers would be relatively easy to level and position. As the height of the pyramid increased, however, they would be more difficult to move into place. To accomplish this, builders used two techniques. Early in the construction process, they levered and dragged stones up ramps, the remains of which have been found at Giza and other sites. As the pyramid grew in height, however, ramps became impractical. If a ramp were used to construct every level of the Great Pyramid, workers would have to pile more material to build the ramp than to build the pyramid. After a certain point in construction, the pyramid itself would be used as a ramp. One of the key differences between the Step Pyramid at Saqqara, and other earlier pyramids, and those at Giza is that at Giza builders incorporated the ramps and staging areas into the pyramid to create smooth faces.

The Egyptian pyramids, and later pyramids built in the New World, were among humankind's most remarkable achievements. Above all, they reveal a social and governmental structure able to coordinate the labor of thousands over extended periods of time. To construct the three pyramids at the Giza site, engineers estimate that a workforce of between twenty and thirty thousand was employed over a period of between eighty and one hundred years. To attribute such an accomplishment as the building of the pyramids to a mystical or supernatural power is to deny the Egyptians and other builders the recognition they deserve. The construction of the world's ancient pyramids is one story that needs no embellishment.

In terms of size and mass, the Great Pyramid and the other Egyptian pyramids were achievements not surpassed until modern times; but, from an engineering standpoint, they were relatively simple structures, piles of carefully shaped and positioned stones. The architectural achievements that would immediately follow did not rival the pyramids in terms of sheer bulk and dimensions, but they demonstrated a growing knowledge of mathematics and a growing understanding of engineering and architectural principles.

From a construction standpoint, walls are relatively easy to build and when in place are fairly stable. Spanning walls to make a roof is another matter. As is true today, in the ancient world there were two basic approaches to spanning: *post-and-lintel* and *arch, vault, and dome.*

In post-and-lintel construction, beams are laid across walls or on top of columns and the space between the beams is in some way filled to make a roof and ceiling. As *Figure 3* shows, this places the top of the beam in what engineers call *compression* and the bottom in *tension.*

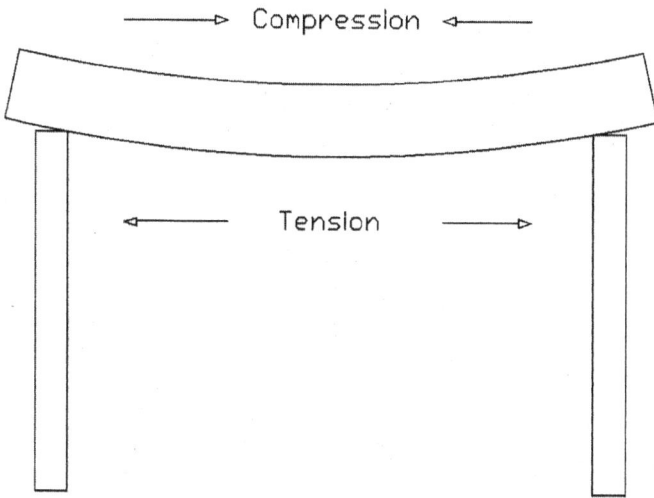

Fig. 3. Tension and Compression. In post-and-lintel construction, beams
are laid across walls or on top of columns. This places the top of the beam in
compression and the bottom in tension.

Wood works well for beams because it has good elasticity and a good balance between tensile and compressive strength—it sags rather than suddenly fails—and even today is used extensively for spanning. Wood, however, rots and burns and for these reasons is not as suitable as stone for monumental architecture. Stone, however, is brittle and weak in tension. As a result, a stone beam can only span a short distance without having to rest on a wall or column. When spanning a large area, like at the temple of Hatshepsut near Thebes, builders have to place a large number of closely spaced interior columns to support the beams. The problem with this is that these columns take up a lot of room.

As *Figure 4* shows, in arch, vault, and dome construction, the spanning element is curved rather than straight.

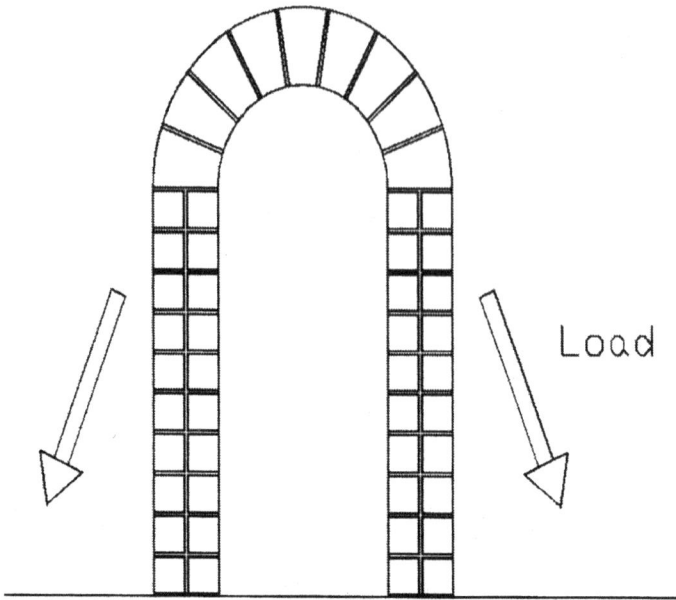

Fig. 4. Arch. In arch, vault, and dome construction, the spanning element is curved rather than straight. The curved shape of the element permits spanning without subjecting the material to tension. This allows stone and masonry, which though weak in tension are strong in compression, to be used to cover large areas.

The curved shape of the element permits spanning without subjecting the material to tension and allows stone and masonry, which though weak in tension are strong in compression, to be used to cover large areas. The forces generated by the weight of the arch, vault, or dome are directed through the walls to the ground. This load, however, is directed outward as well as down, which requires walls that are massive enough to carry the load or walls that are in some way buttressed, *Figure 5.*

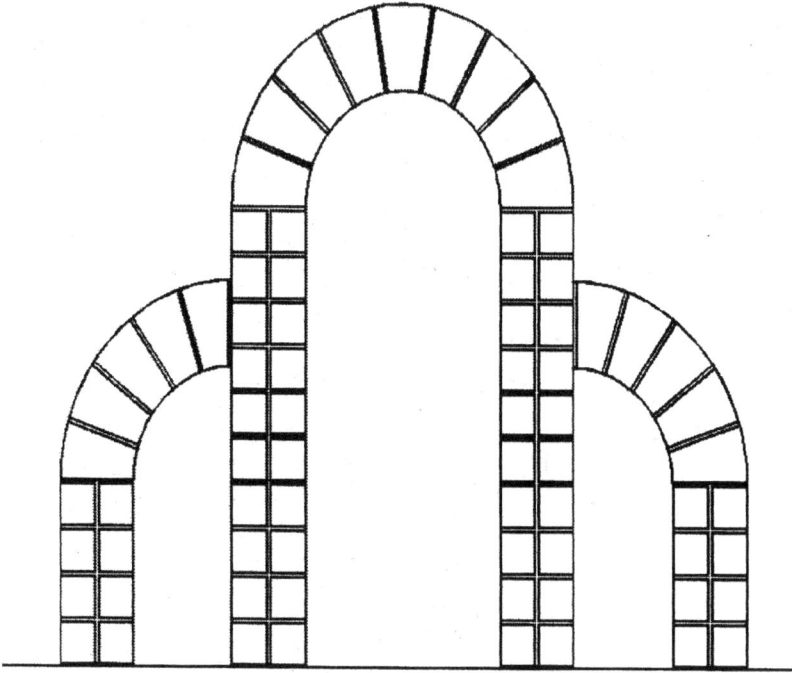

Fig. 5. Buttress. In arch, vault, and dome construction, the load is directed out as well as down. This requires walls that are massive enough to carry the load or that are in some way buttressed.

In the earliest construction, roofs were spanned almost entirely with wooden beams run between walls. To build a simple shelter such as those constructed in rural Africa today, the builder erects a circular or rectangular wall frame made of thin vertical poles, spans it with roof poles that rise to a peak or ridge, and ties everything together. He then plasters the walls with mud and thatches the roof with grass. In later, monumental construction, roofs were spanned with stone beams. Arch, vault, and dome construction was used in a primitive way as far back as Ancient Egypt, used in clay block structures in Ancient Persia, further experimented with in Ancient Greece, and developed into a standard building form in Ancient Rome.

This brings us to Ancient Rome's most important contribution to engineering. Not only did Roman architects and engineers make extensive use of arch, vault, and dome construction, the Roman builder was responsible for a discovery that would forever change the face of the urban landscape, an innovation that would dominate building technology since its advent and that today is a cornerstone of construction. The Romans invented concrete.

At some point in time, Roman builders discovered that if they heated limestone to the point where it crumbled into a powder and if, when cool, they mixed this powder with water it would get hot and solidify, binding sand and gravel in a stone matrix. Though crude, the cement that held together this mixture, called *quicklime*, was strong enough to mortar the stones of the coliseum and Roman aqueducts. Builders also used it to cast the massive concrete dome of the Pantheon. Completed in the year 128, the rotunda dome measures 43.2 meters, or 142 feet, in diameter and has an 8.5 meter, or 28 foot, oculus.

With the fall of Rome, Western architecture evolved through the Christian, Byzantine, Romanesque, and Gothic periods. Just as Greek and Roman buildings lacked the massiveness of the pyramids but embodied a higher level of engineering, the buildings of the medieval world were often not as imposing as earlier structures but over time showed a refinement of construction skills.

They also showed a refinement of architectural principles. Unlike the engineer, who is concerned with the function and structural integrity of the building, the architect is concerned with its appearance and artistic statement. Throughout time, certain mathematical principles have guided architects in their pursuit of beauty. The simplest is the principle of *symmetry*. A pyramid, for example, is symmetrical. Each face is the same and slopes away at the same angle. Closely related to the principle of symmetry is that of *repetition*. Often in architecture, a certain feature, an arch for example, is repeated. Portions of the Roman aqueducts are supported by lines of simple vaults. Roman buildings were often constructed with walls made of lines of large arches that supported walls made of lines of smaller arches. As this example also illustrates, dimensions that double or by some other formula systematically increase in size are also important—large arches that are say twice the width and height of small arches.

Around the year 1200, the Italian mathematician Leonardo de Fibonacci came across a sophisticated variation of this principle, the *Fibonacci series*. In this recurrent series, $k_n = k_{n-1} + k_{n-2}$, each number is the sum of the two numbers that preceded it: 1, 1, 2, 3, 5, 8, 13, 21, and so forth. Like a simplistic representation of evolution, the series builds on the old to create the new. And, as we would expect from such a representation of evolution, its proportions are found in nature. The curve of a nautilus shell spirals outward in a Fibonacci pattern. The growth patterns on the leaves of certain trees spiral outward in a Fibonacci pattern. The dimensions of the human body can be interpreted to represent a Fibonacci pattern. Architects noticed something else about the Fibonacci series. Many of architecture's traditional and most pleasing forms followed this pattern. The ratio of successive terms in the Fibonacci series converges on what architects call the "Golden Ratio." When proportioned according to the Fibonacci series and golden ratio, columns, furniture, and buildings looked right, natural, aesthetic, pleasing to the eye.

The principles of symmetry and repetition, and of the Fibonacci and other series, are so fundamental to architecture that they influence design to this day. The twentieth century American architect Frank Lloyd Wright was well known for basing his designs on Fibonacci proportions.

From our earliest settlements to the pyramids of Egypt, to the cities of ancient Greece and Rome, to the Medieval cathedrals, to the wide streets and boulevards and soaring domed churches of the Renaissance—in the architectural traditions of China, India, Russia, North Africa, and the New World—our urban environment reflected our evolution. As we grew to better understand our world and to create less collective social orders, we molded our urban landscape in ways we deemed practical and beautiful.

At the end of the eighteenth century, the industrial revolution swept across Europe, and the design of our urban landscape entered its most recent phase. Locked in a mechanistic view of the world, architecture and urban planning lost much of the human quality prevalent in colonial and earlier periods and became increasingly utilitarian. Buildings were square and practical, their architecture driven less by aesthetic considerations than by cost and construction time. In 1811, New York City planners divided Manhattan Island into identical rectangular blocks to encourage access and settlement. In our quest to partition the earth in an orderly way, we stretched roads and tracks across Asia, Africa, Europe, and North America. Our environment reflected the growing importance in our lives of commerce and material wealth.

Our return to recent times also returns our discussion to the countryside, where until the early twentieth century our agricultural practices had advanced remarkably little since the advent of cultivation. For millennia we had worked the land with our hands and, when fortunate, with animal labor. Water was provided by rain and local irrigation systems. For fertilizer, we relied on manures and human waste.

When we lived as nomadic hunters and gatherers, we enjoyed a varied, protein rich diet, which contributed to longevity, tall statures, and general good health. With urbanization and agriculture, diets became less varied and included less meat, fish, and other sources of protein. In the days of seignorialism,[3] peasants ate almost no meat or other primary protein; and, depending on region, diet was dominated by a single food crop: typically rice, wheat, or potatoes. The diminutive statures, short life spans, and general poor health of people living during the Middle Ages is well documented and may have contributed to the number of deaths experienced during plague times. It would not be until the early twentieth century and what is popularly called the "green" revolution that diets would improve.

The green revolution centered on scientific practices that led to the breeding of more productive strains of rice, wheat, and other food crops. To realize the yields genetically possible in these crops, however, they had to be grown in rich soils and required greater amounts of water. Greater food yields also made crops more susceptible to insects. Advances in chemistry led to pesticides. Dams and irrigation systems such as, in the United States, the Hoover Dam on the Colorado River and the Tennessee Valley Authority's system of irrigation and hydroelectric projects provided the water. Advances in chemistry also led to chemical fertilizers that allowed soil to be enriched on a scale not possible by the use of manures. Chemical fertilizers contain the three key sources of plant nutrition—nitrogen, phosphorous, and potassium—and lesser sources such as sulfur, calcium, and magnesium.

In the developed world, where the resources were available to build dams and irrigation systems and to synthesize pesticides and fertilizers, the advances of the green revolution allowed agricultural production to soar. Diets became more varied and, with abundant feed for livestock, higher in protein. In certain respects and among certain populations, we ate what our hunting and gathering forbearers ate, the diet we were drawn to and which our bodies had over millions of years evolved to digest and make the best use of—a varied diet that included ample amounts of meat

3. See *Book 2, Economics of Fulfillment*, chapter 3.

and other protein. Many factors contributed to better human health and increased longevity during the twentieth century. A better diet, sewage disposal, and medical advances are considered the most important; and, arguably, nothing has done more to elevate the human condition than the simple chlorination of drinking water.

One other factor contributed to the green revolution, one that not only made highly productive agriculture possible but one that marked a key turning point in humankind's molding of the earth's reflective landscape. At the end of the 1800s, the internal combustion engine became a practical alternative to the steam engine. Small and lightweight, it made possible the mechanization of farming and construction. It also made possible a development that would alter the earth's urban and ecological infrastructure in ways scarcely imaginable decades earlier. In 1908, Henry Ford began mass production of an article known, loved, and hated by all—the automobile.

4

Roundness of the Earth

To grapple with the present state of our planet's landscape and to appreciate the significance of the automobile in defining that landscape, we begin on a philosophical note. Among Teilhard de Chardin's many qualities was the skill with which he could sum up his conclusions in a simple thought and express that thought in a way we could picture and come to terms with. One of his most striking visual metaphors is a concept called *roundness of the earth*. Given our nature as creative beings, and the drive to extend our influence our nature inspires, Teilhard saw that at some point in time humanity must draw in on itself. Ideas and cultures cannot expand outward as if on an infinite plane. The roundness of the earth brings us together; it intensifies our thoughts and actions. Where has this focusing of human consciousness carried us? What is the present state of the spherical body we call home?

With regard to the earth's landscape and humanity's interaction with the earth's landscape, two opposing forces have come to, at present, shape our rural and urban environment. At one end of the spectrum are *economic interests*, factions driven by values of profit, exploitation, and development. On the other end are *environmental interests*, factions driven by values of regulation, restoration, and preservation.

To understand this dichotomy, we begin with a review of certain points developed in *Threshold to Meaning: Book 2, Economics of Fulfillment*. As we discussed in that book, and touched on in the previous chapter, prior to urbanization twelve thousand years ago, humankind lived as nomadic hunters and gatherers. As such, we interacted with the biosphere much as any other species in an ecosystem. We enjoyed a place in the ecosystem, and the ecosystem gave us the material resources we needed to fulfill our role. With urbanization, we abandoned our ecological niche. Only a limited amount of game could be hunted and fruits and tubers gathered within a reasonable distance from a settlement. Because of this, the community experienced a chronic shortage of food and other items. It existed in a state of intrinsic *scarcity of resources*.[1]

Scarcity was a general condition in the early urban community; and it affected our lives in the most profound way. In the settlement, we consumed resources in a different place from where we acquired them. As a result, we had to devise systems to coordinate the way we gathered resources in the countryside, where they existed in surplus, and to divide them up in the village, where they existed in deficit. Urbanization founded economics as we typically think of it today—*scarcity-based* economics.

Urbanization and the condition of scarcity also changed our view of the world. For the first time, we did not see ourselves as part of our environment but separate from and in opposition to our environment. Our lives were a struggle to eke out of the earth what we needed to feed, cloth, and shelter ourselves and our community. The earth was our adversary, and we had to take from it what we needed and desired. Urbanization and scarcity nurtured humanity's ideals of wealth and property, of survival, competition, and economically based social structure. Urbanization and scarcity also brought into human awareness the concept of nature—the idea that we are separate from the natural world.

This scarcity-based interpretation of the universe underlies contemporary economic and environmental views. Both ideologies rest on the assumption that material resources are by nature limited and human material wants are by nature unlimited. As such, both rest on the notion that man is in opposition to nature, however we may at the moment define nature.

With regard to economics—a topic that in light of the earlier book we will only bring to a point—two philosophies form the basis of our global economy: socialism and capitalism. In this regard, every nation employs a blend of socialistic and capitalistic practices: socialism-capitalism. In most nations, however, capitalism has emerged as the dominate ideology, or at least as the ideology that is largely responsible for economic prosperity.[2]

---

1. For a study of the origins of scarcity-based economics see *Book 2, Economics of Fulfillment*, chapters 2 and 3.

2. Although our discussion centers on capitalism, please note that socialist nations—exemplified by the rapid industrialization of the former Soviet Union and its almost nonexistent health, safety, and environmental practices—can be as exploitive of resources and the environment.

In the capitalist sphere, our economic practices are motivated by profit and justified by values of survival and competition. Reflective of the extent to which we embrace these values is money. To us, today, money is everything. It is that which we desire and that for which we harness our creative power to obtain. Money is the carrot that motivates us in life. The lack of money and the threat to survival the lack of money represents is the stick. We have created an artificial object of value, money, and we have placed it at the center of all that we do and dream.

The value we place on money tells us something about today's economy and our interaction with the economy. Whether socialistic, capitalistic, or their variations, we embrace contemporary economic theories and practices with blind devotion. Faced with problems under capitalism we look for solutions in socialism. Faced with problems under socialism we look for solutions in capitalism. When have we challenged economic ideology itself? When have we confronted the applicability of our materialistic worldview, the notion of scarcity?

As many have written, and as in *Book 2, Economics of Fulfillment,* we have provided the background to further support, in today's world economics is more than a philosophy of wealth creation and distribution and of the mechanisms through which the production and exchange of goods and services takes place. Economics is a dogma, a canon, a doctrine. We worship at the altar of the global economic machine. To us, today, economics is a religion.[3]

On the opposite end of the ideological spectrum, we embrace environmentalism with no less conviction. The environmental movement began in the mid-nineteenth century as a reaction to the massacre and near extinction of the bison herds that once roamed the Western United States in numbers too large to comprehend.

In 1845, the movement was given voice by the American writer and naturalist Henry David Thoreau. Influenced by the transcendentalist philosopher Ralph Waldo Emerson, Thoreau abandoned civilization to live in a crude hut built on the shores of Walden Pond, a lake on the outskirts of Concord Massachusetts. Though his time in the woods was not as reclusive as he proclaimed, Thoreau's exploits did help to instill into the national mindset the idea that nature was more than a source of timber

---

3. For a better development and justification of this conclusion, see *Book 2, Economics of Fulfillment,* chapter 4.

and minerals. Thoreau saw in the natural world a spiritual value, an opportunity for humanity to commune with the earth and with itself.

The naturalist George Perkins Marsh furthered this idea in his 1864 book *Man and Nature*, where he brought into human consciousness the notion that we were making permanent and disastrous changes in the global environment. He also put forth a more radical concept. In Marsh's view, we should allow disturbed environments to heal on their own or we should restore them to their natural state.

In the late 1800s, the views of Marsh, Thoreau, and others in the early environmental movement led to the establishment of the first national parks and influenced the outdoorsman and President Theodore Roosevelt, who, in the early 1900s, set aside 125 million acres as national forests. The primary objective of Roosevelt and of Gifford Pinchot, the first chief of the United States Forest Service, however, was not esoteric. It was to preserve tracts of timber, grassland, and mineral reserves for future use and development.

There were others, however, who embraced the preservation of natural places for less practical reasons. Among them was the explorer and naturalist John Muir, who in 1892 founded the Sierra Club and who felt that natural areas should be left undisturbed to meet our aesthetic and spiritual needs. Muir also saw a value in preserving natural areas to meet the aesthetic and spiritual needs of future generations. We have a duty to our children, Muir felt. We have an obligation to preserve our environment so that decades and centuries from now we can experience the natural world.

The dust bowl of the 1930s and other environmental disasters prompted President Franklin Delano Roosevelt to establish the Soil Conservation Service and the Civilian Conservation Corps. These groups built trails and lodges, reclaimed farmland, and restored damaged ecosystems. Although the early environmental movement was centered in the United States, it touched many parts of the world. In Africa, hunters who feared that a loss of wildlife habitat would cause the game species to go the way of the American bison drove the creation of game reserves, including, in 1951, Tanzania's Serengeti National Park.

The modern environmental movement began in 1962 with the publication of Rachel Carson's book *Silent Spring*. In this book, Carson argued that setting aside wilderness areas and wildlife refuges was not enough to maintain a healthy environment. She established a link between resources, pollution, and human health. In response, congress passed the Environmental Educational Act, the National Environmental Protection Act, the Wilderness Act, and the Endangered Species Act. It also established the Environmental Protection Agency to monitor the standards set by the Clean Air and Clean Water acts. In Western Europe, nations enacted similar legislation; and, in 1975, the United Nations began a program to promote environmental education worldwide.

Even the staunchest critic of the environmental movement must concede that environmentalism has given us much. Without environmental activism and the populist pressures it has brought to bear in mediating the profit mechanism of scarcity-based economics, we would not enjoy the quality of air and water and thus of life that at least in the world's developed nations we take for granted. But the environmental movement of today is not the movement of Muir, Marsh, Carson, Thoreau, or the Roosevelts. As one who has spent much of his life outdoors and deeply respects the natural world, it is not the movement I grew up to admire.

In this respect, environmentalism has followed a familiar developmental path, one that tells us much about the present movement. Faced with an issue, we form organizations to pool our resources and political power. As the environmental movement of the 1960s and 1970s shows us, our organizations can allow us to further our agendas. But whether it is a business, government, or environmental organization, when the organization begins to meet the goals it was created to achieve, the focus of the organization invariably shifts. Whereas one might think that an organization would define a new set of goals and channel its energy to their obtainment, it almost always channels its energy to furthering the organization itself. To accomplish this, the organization replaces practical objectives with ideology and uses that ideology to shift its efforts to propagation and to building financial and political support. The organization becomes more important than its goals, which are defined to achieve political rather than practical ends. The Sierra Club wages all-out war on the sport utility vehicle, or SUV, but pays little attention to high-sulfur coal burning in Asia, a monumental health and environmental concern. Other groups fight the flush toilet and the use of not only disposable diapers but of all diapers. The environmental community has transformed itself from one driven by an altruistic agenda, centered on creating a better place for humanity, to one driven by doctrine and ideology.

Central to the environmental system of beliefs is a conception of nature derived from the idea of the ecosystem as it was popularly defined in the 1960s, a perspective where systems are seen as intricately functioning mechanisms that in their ideal state are in balance. As such, the biosphere is interpreted to be a vast machine where, prior to the ascent of civilization, every species interacted harmoniously with every other species—a complex, in-sync contrivance in an idyllic state of equilibrium.

The biosphere, however, has never been in a sustained state of equilibrium. It has been in a state of evolution. Moreover, as we established in chapter 2, *Pre-Reflective Landscape,* since the advent of reflection and, with sweeps from the simple to the complex along the way, as far back as the mass extinctions at the end of the Paleozoic era, the net direction of the biosphere's evolution has been toward decline, toward a decrease in complexity and a reduction in the number of species and ecosystems. Paleontologists tell us that in excess of ninety-nine percent of all species that have ever lived have fallen into extinction. Unwilling to incorporate paleontology into their worldview, the environmental movement takes humanity to be that which has destroyed an idealized conception of nature. As such, we are the enemy. We are that which to save the deity of the earth we must confront and in the end defeat.

From this philosophical base, groups in the United States have proclaimed goals as far reaching as depopulating the rural West and reintroducing the grizzly bear to habitats in the lower forty-eight states, a move that would effectively close many of the West's wilderness areas to the casual hiker and backpacker. Other groups have burned ski lodges and lumber mills and fire bombed sport utility vehicles. Still more extreme, groups have called for eliminating meat from the human diet because cattle and other stock release methane which, arguably, contributes to global warming.[4] Other environmentalists have called for the planned extinction of the human species.

At the 1992 Earth Summit in Rio de Janeiro, the United Nations adopted "Agenda 21," or guidelines for "sustainable development." The core assumption on which Agenda 21 is based is the belief that the human being has no greater place in nature than any other species. Based on this ideal, groups have adopted goals that have ranged from those of clear benefit to us and the environment, such as improving third-world economies, to restrictions on human activity more severe than any George Orwell or

4. See chapter 10, *Climate.*

the most ruthless communist dictator could have envisioned. In the more radical of these schemes, a global government would oversee a reduction in the earth's population from today's six billion to a few hundred million. Human habitation would be limited to "green" communities, and the majority of the planet would be off-limits to human entry. The government would eliminate private property, ration water and energy, and control housing, industry, agriculture, and transportation. Pregnancy would require government approval. Education would be government sanctioned and environmentally based.

There are well-intended members of the environmental community, and the most extreme views and actions do not represent the movement as a whole. But the movement, as a whole, has become remarkably dogmatic. The aim of today's environmentalist has become to turn back the clock, to return the earth to the way it was before "thrown out of whack" by the materialistic cravings of humankind. Muir, Marsh, Carson, Thoreau, and the Roosevelts saw the environment as inseparable from the human being, as nurturing to the spirit and necessary for our future. During his years in California's Sierra Nevada Mountains, John Muir made his living operating a sawmill. Today's environmentalist sees the human being as irreconcilable with the environment. We are the enemy. We are to be controlled and defeated. With no less conviction than that felt by the capitalist who lusts for profit, environmentalists embrace the notion of man against an idealized image of nature. The objective of environmentalism has become political and ideological. Environmentalism, today, is a religion, its goal to return the earth to its "natural" state—to the form "God had intended."

On one end of the exploitation-environmentalism spectrum, we have factions driven by a philosophy of profit and development embraced with religious conviction. On the other end, we have factions driven by a philosophy of stagnation and conservation embraced with religious conviction. Both views are an outgrowth of the human move to urbanization. Both rest on an obsolete scarcity-based interpretation of the human being and its place in the world.

The polarity in environmental management is clearly visible in today's rural and urban landscape. The earth's landscape consists of four major divisions. There are large tracts of *wilderness* land, areas where ecosystems function much as they have for millennia. Untouched land remains in Alaska, Siberia, the Himalayas, the Canadian Northwest

Territories, the outback regions of Australia, and the rain forests of Africa and South America. Antarctica is almost entirely untouched, and smaller wilderness areas are found in almost every nation. Bordering wilderness areas are vast tracts of land that retain many natural characteristics but that we routinely enter for mining, grazing, logging, and recreation. These *intermediate zones* are accessible but not closely managed or heavily developed. Nestled in the earth's intermediate zones are regions of intense human intervention: *agricultural land* and, even more intensely managed, *urban land.*

To illustrate the conflict that exists between economic and environmental interests, we begin with a look at the management of our rural landscape and draw on an example from the western United States. I chose this example because I have experience in the region. Parallels can be drawn to ecological systems throughout the world. In that we delved into the limitations of scarcity-based economics in *Book 2, Economics of Fulfillment,* here again, we will weigh our discussion toward limitations on the environmentalist end of the spectrum.

Fire is a natural component of a forest ecosystem. In areas of the West that get a substantial amount of rainfall or are at a high elevation, the fire cycle, or the length of time between burns, may be a century or more. In dryer areas and areas at lower elevations, the fire cycle may be as short as a few years. Prior to settlement, Native Americans routinely burned large tracts of forest. The open woodlands of large diameter fir, pine, cedar, and spruce penetrated by Lewis and Clark in the early 1800s, and that they took to be in a natural state, were in many areas the result of thousands of years of human management.

Settlement and the relocation of native peoples ended the practice of burning. Forests with a density of 40 to 50 trees per acre increased to densities similar to those we see today, as high as 2,000 trees per acre. This situation climaxed with the massive fires of 1910, which consumed an estimated 20 million acres. As a result of these fires and of the loss of lives, homes, and timber they brought, we came to see fire as our enemy. This view, compounded by our ignorance of the role of fire in the forest ecosystem, led to extensive fire control.

Fire serves many purposes in the forest ecosystem. Heat allows the seeds of trees such as the lodgepole pine and the giant sequoia to open and germinate. Periodic burning controls disease, promotes diverse vegetation and animal life, and prevents brush and small trees from growing to a height where they spread flames into the forest canopy. When we began fire control, we disrupted the fire cycle as it took place in nature and as it for ten millennia took place at the hands of Native Americans. This had the greatest impact in dry, low elevation areas where fire was a frequent event. Fuel loads increased. Diseases and insect infestations spread. Plant and animal diversity decreased. As recent fires in California and the Pacific Northwest, including Oregon's 500 thousand acre Biscuit fire, illustrate, brush and other fuels have built up in Western forests to such an extent that when fires do start they burn so hot and grow to such size that they kill large, fire-resistant trees and are difficult to contain. They demonstrate fire behavior not seen in a forest where fire has been allowed to periodically clean things up.

This brings us to the issue of forest management. When we peel away the layers of dogma and politics, two points stand out. First, to maintain forest health, we must reintroduce fire into the forest ecosystem. Fire is an aspect of a forest ecosystem, and we cannot do without it. Second, unless we take ourselves out of the equation—depopulate the rural West as some have interpreted United Nations Agenda 21 to dictate—we cannot reintroduce fire in a natural way. At our present moment in evolution, humanity has spread across the Western United States to such an extent that it is not possible or reasonable to remove us as a factor in forest management and let "nature" take care of itself. We must reintroduce fire into the forest ecosystem, but we must do so in a way we can live with. There are those who will not accept this conclusion, but from the standpoint of common sense its derivation is unavoidable. We are here, and we are going to stay.

As obvious as the above points may be, foresters are blocked in their efforts to manage our forests by the conflict that rages between economic and environmental interests. Members of the timber industry lobby to log the large, fire-resistant, and most profitable trees. Members of the environmental community lobby to stop all logging and to end all human intervention—thinning, reforestation, salvage logging, and controlled burns. In their attempts at ecosystem management, foresters face political battle after political battle, legal ruling after legal ruling.

Rural lands throughout the West are caught in a struggle between ideologies. Ranchers in Idaho face reintroduction of the wolf and annual losses of sheep and cattle that number in the thousands. Ranchers in Wyoming face the spread of the grizzly outside of Yellowstone National Park and the threat to safety and livestock it poses. Farmers in Oregon face environmentalists who, brandishing the sword of the Endangered Species Act, are bent on the demolition of dams on the Klamath River and on the dismantling of irrigation systems that have supported a vital agricultural economy for a century. Bogged down in legal bickering, state and federal forest lands face fire, brush, and disease, while the best managed private forest lands have integrated timber harvest into the forest ecosystem to such an extent that they not only maximize production but have become models of ecosystem health reminiscent of forests that existed centuries ago. Reason, science, and common sense are powerless in a battle we wage between dogmas.

A similar situation exists in the management of our urban landscape. Today's cities embody a blend of new and old designs imposed on an urban model inherited from the days of horse-and-buggy transportation. And, like cities throughout history, those of today reflect our understanding of the world. They embody our economic and environmental ideologies.

One characteristic more than any other defines the modern urban environment. For the most part, today's cities are geographically disperse. Population density is low and spread over a large geographical area. This uniformity has many consequences. Most apparent, it creates a situation where the individual has no clear way to get from one center of human activity in the city to another, such that centers of human activity exist. Modern urban design incorporates a *disperse* transportation system. In this system, we travel directly from any one geographical location to any other geographical location. If we were to think of the city as an electrical circuit, it would have wires running from every terminal to every other terminal.

And, like an electrical circuit with wires running from every terminal to every other terminal, the transportation system of the modern city often shorts out. The epitome of contemporary urban planning is Southern California's Los Angeles basin. Bounded on the west by the Pacific Ocean and by mountains on the north and east, the Los Angeles metropolitan region has a population in excess of sixteen million sprawled over a vast geographical area. So spread out is the basin's population that the only practical way to get around is by car, in particular if one has to travel a distance. Every building in the urban area is linked to every other building by a network of roads. This situation creates the snarled freeways and clogged city streets for which Los Angeles is as famous as it is for its motion picture industry. The Los Angeles basin represents an extreme of modern urban design, but every city faces similar planning issues.

Given the absurdity of using a vehicle that can weigh more than two tons—as heavy as the stone blocks used to build the Great Pyramid—to move a person that can weigh less than one-hundred-and-fifty pounds from every point in a city directly to every other point, attempts have been made to improve urban design. Here we find conflict between business and environmental interests and the environmental and land-use-planning failures that result.

As any contractor knows, the way we build our homes and cities has as much to do with politics as with engineering and architecture. Via the power of government, planners force our building and land use activities to conform to political values in two ways. First, they set mandates and standards, or apply methods of direct regulation. Second, they control the cost and availability of goods and services, or manipulate the market. Both approaches have been used with some success. In the United States, the air and water are cleaner than they were three decades ago. Both approaches, however, have also had unintended consequences.

Mandated by government regulation, many cities in the 1960s and 1970s established urban growth boundaries to prevent sprawl. As a result, many small homebuilders, those who took pride in their craft and who sincerely wanted to better their communities, could not afford permits and system development fees, much less the delays and legal costs incurred to overcome the land-use planning bureaucracy. This left in business the large, highly capitalized developers—firms that mass-produced homes and neighborhoods. In some areas, urban growth boundaries have limited sprawl. They have also driven up the cost of construction, lowered the quality of homes and other buildings, and, contrary to intent, encouraged subdivisions, strip malls, and other capital intensive projects. In a state such as Oregon, where only about three percent of the land falls within designated urban boundaries, one must ask if the tangible, as opposed to the political, benefits of limiting sprawl have exceeded the costs.

The primary tool regulators use to maintain urban boundaries and to restrict property use is zoning. No one wants to live next to an oil refinery or across the street from a coal-fired power plant. No one wants to live downwind from a steel mill or in the flight-path of an airport. Zoning laws help prevent one property owner, through his or her use of the land, from imposing on the property rights of other property owners. In this respect, zoning is essential; coherence in urban design requires planning. Invariably, however, legislators take it too far. The needless separation of business and residential land has forced commuters to travel from their homes to distant commercial areas and increased our need for energy, highways, parking structures, and automotive support facilities. In some communities, the simple removal of a tree, a change of house color, or a minimal change in landscaping requires a costly permit, approval, and inspections. Historic districts have popped up everywhere. Certain structures and neighborhoods have historic value. But preserving something old for no other reason than it is old is not historic preservation. Horror stories of arbitrary historical standards and fanatical inspectors abound to such an extent that many contractors refuse work in historical districts.

The regulatory approach to compelling politically sanctioned decisions has had other unintended consequences. As we mentioned, the Sierra Club wages an all-out war on the sport utility vehicle. If you listen to the environmentalist party line, you would think that every social and economic problem the world has or will ever face is the result of our choice to drive an SUV. What the propagandists behind this battle fail to tell us is that the popularity of the SUV was the outcome of overzealous energy and environmental regulation. Motivated by an OPEC oil embargo and lines at the pump, President Jimmy Carter drove through energy conservation legislation in the late 1970s that required cars to meet strict gas-mileage standards. This forced out of production the vehicle that for a generation had been the staple of American suburbia, the full-sized station wagon. In need of a car that could carry kids, pets, and sports equipment and pull a boat or trailer, families had no choice but to turn to the SUV. Classified as trucks and thus exempt from strict mileage requirements, they offered the suburban family the large vehicles they required, albeit vehicles that in many respects were less safe and efficient than the old station wagon.

Government also regulates our environmental behavior by manipulating the market forces of capitalism. In the 1970s, a widely held belief was that environmental and land-use problems were the result of an improperly functioning economy. Specifically, economists felt that supply and demand did not assign the "correct" price to commodities.

Gasoline, for example, should not be priced based on the cost of crude oil, refining, the supply available, and the amount people are willing to pay but should include the cost of wildlife habitat lost to oil facilities and road construction and the cost of cancer, emphysema, and decreased quality of life caused by air pollution. If government added these costs to the price of gasoline by imposing a tax, gasoline would be more expensive. Demand would drop. We would drive less, and the automobile would have less environmental impact.

Recent variations on this approach include the *carbon tax*, a tax on all forms of energy proposed in the United States under the Clinton-Gore Administration and today sought by the United Nations as a global tax levied against citizens of that and other developed nations, and the *carbon offset*, a reworking of an idea popularly called "pollution credits."

A legal tool used to enable coal-fired power plants and other industrial facilities that do not meet current emission standards to remain in operation, pollution credits allow facilities that pollute more than a governmentally determined amount to purchase credits from facilities that pollute less and offset their emissions. In the carbon offset scheme, those of us who exceed a governmentally determined standard for carbon emissions, which equates to a ration of energy, buy credits from those of us who do not. If we were to buy an SUV that fails to meet a fuel efficiency standard, for example, we would, at the time of the purchase as is now done in Great Britain, with each trip to the service station, or through some other method, also have to buy a carbon credit to compensate for our additional energy use.

We will address the issue of climate change and the theory of global warming and greenhouse gas emissions in a later chapter. Beneath the environmental hype, however, the carbon schemes—labeled tax, cap and trade, or otherwise—do nothing other than increase the cost of energy with the intent to reduce its use.

As the theory of market manipulation goes, if we used the tax system or some other legal tool to adjust the price of every good and service to the "proper" amount, economic activity would decline to the point where our streets were no longer crowded and our industries and other activities no longer had a significant impact on the environment. Capitalism and environmentalism would work together, in harmony. Man and nature would be in balance.

But how do we establish the price of an indirect cost such as those associated with the automobile and the use of gasoline, or of other even less tangible costs? As every environmental economics professor asked their students in the 1970s: "What is the dollar value of a sunset?" More important, how do we maintain a reasonable standard of living in a shrinking economy, and how do we provide for development in the Third World? As we discussed in *Book 2, Economics of Fulfillment*, contemporary economists disregard a static supply-and-demand model of economic behavior in favor of a dynamic model centered on the capital cycle. There is no shortage of politicians though, who, their heart in the seventies and their political base in the environmental community, call for higher gas taxes and for regulations controlling virtually every aspect of life.

The city planners of today manage a disperse, automobile based urban environment; and, aware of the problems with that environment, they have for political reasons, and one hopes a belief in the good of their decisions, attempted to restructure it in a way where we are less dependent on the car. The result of their methods of direct regulation and market manipulation, however, has been a dysfunctional automobile based urban environment.

The so called "green" city of Portland in the state of Oregon is a good example. Anti-automobile policies have limited lot size to increase population density and funded bike paths at the expense of highways. Drivers parked on clogged freeways now have a good view of empty bike lanes while idling engines gobble fuel. In an attempt to integrate business and residential space, Portland and other cities have encouraged "planned-unit developments," or communities designed to integrate work and home life. Those who take the minimum wage, coffee-shop jobs people want near their homes in these typically upscale communities, and that are most often provided for in the planning process, however, cannot afford to live in these developments. Those who can commute to high paying jobs elsewhere.

We have created an automobile based urban environment where it has become harder and harder to get around by car but where public transit offers no practical alternative. How, in a usable way, can we link every geographical area to every other geographical area by light rail? So dysfunctional is mass transit in our present urban model that in New York City—the urban center in the United States that relies most heavily on public transit—only about twenty percent of commuters regularly take the bus, train, or subway. In the transit-intensive city of San Francisco, about seven percent of commuters rely on public transportation. In the typical United States city, the figure runs at between one and three percent. Light rail systems have proven to be particularly inefficient. The cost to the user to take the train may be inexpensive, but the cost to the taxpayer is often exorbitant. The subsidized cost per light rail rider in Portland and similar cities is by some estimates as high as $16,000 per year. Critics point out that it would be cheaper to lease every rider a new luxury sedan or SUV and pay for his or her gas and insurance.

Though often held up as a model for the world, European cities face their own planning challenges. Whereas American cities have been built from the ground up over the last two or three centuries, European cities have evolved out of an urban framework that in some instances has been in place for two or three millennia, a framework that not only predates the automobile but the common use of the horse. This has led to an urban layout with higher population densities than in the United States and has made commuting by foot and bike more practical. Still, though, there is no coherent way to travel distances in the city. Urban mass transit systems in Europe face the same geographic limitations as those in the United States and are slow and circuitous. Moreover, traffic patterns in cities such as Rome and Paris, which has considered banning cars altogether, are so snarled few Americans would tolerate them.

As urban planners, we face a choice. We either accept the car for what it is and design a workable automobile based urban model or, as we will do in the book's next section, accept the obsolescence of such a model and, rather than try to patch it up through politically motivated energy and environmental policies, invent a new urban model.

The earth's rural and urban landscape mirrors our state of mind. Our cities and countryside reflect our uncertainty, our absence of a greater understanding of the universe and of the sense of order such wisdom provides. The individual often feels a lack of community, and economic factors often compel family members to live great distances apart. We feel alone, isolated, ill-suited to cope with a world rushing to some undefined future. Contemporary rural and urban planning reflect a global community that has only begun to rise from the collective social orders of the past and embrace the means to coordinate engineering on a planet-wide scale.

As a human community, we have yet to come to terms with the roundness of the earth. Symbolic of our worldview is that which we have spoken about in the context of the city, the automobile. On one hand, the automobile is a machine, an object that transforms chemical energy into kinetic energy and that allows us to move with greater speed and comfort. On the other hand, the automobile is a symbol, a metaphor. It is the embodiment of our desire to extend our influence outward, an icon reflective of the days when we reached across the earth's surface as if it went on without end. The automobile dramatizes our uncertainty, our reluctance to grapple with where we are going and with where we want to be. The automobile encompasses our failure to see the earth from afar, our unwillingness to accept our planet's spherical geometry.

Four billion years ago, the universe entered the evolutionary period of life. One hundred thousand years ago, the universe entered the evolutionary period of understanding. With this transcendence, the biosphere fell into the realm of evolution's trailing arrow, and the leading arrow of creation locked in the human quest to comprehend itself and its world. With the advent of urbanization and agriculture, we for the first time played a conscious role in the biosphere's devolution. Today, as humankind crosses the threshold to meaning, and as we setout to draft our blueprint for reconstruction, this role transforms. The roundness of the earth—our insight into our planet's post-reflective landscape—draws us in on ourselves. Our understanding carries us beyond contemporary economic and environmental doctrine and compels us to embrace a new approach to urban and ecological management.

Contemporary economics drives us blindly forward in a quest to harness the earth's resources. Contemporary environmentalism drives us blindly back in a quest to restore the biosphere to some imagined state of environmental correctness. Both views are the product of centuries old reasoning. Both are the outcome of an obsolete, scarcity-based interpretation of the world and a man-as-separate-from-nature ideal. As architects of the future, our role is not to exploit the earth. Neither is it to roll back the clock and return the earth to a past evolutionary state. To engineer the landscape of tomorrow, we must move beyond notions of man and nature, of economics and environmentalism and elevate ourselves to a new relationship with our reflective landscape. As we cross the universe's threshold to meaning, we take upon ourselves the goal to manage the biosphere in a way that meets our needs as evolving beings in an evolving universe. We take upon ourselves the objective to reconstruct the earth's urban and ecological infrastructure in light of the universe's origin, evolution, and future. Humanity defines the leading edge of the creative process. As such, the trailing edge of creation is ours to direct. Our gaze locked on the future, our consciousness wrapped around all time past, we must put behind us contemporary economic and environmental dogma and take command of evolution's trailing arrow.[5]

5. See *Book 1, Evolution of Consciousness*, chapters 8, 9, 10, and 14.

PART TWO

Blueprint for Tomorrow

5

Design Criteria

HUMANITY'S TRANSCENDENCE TO MEANING and evolution beyond scarcity-based economics empowers us to reshape our surroundings in almost any way we imagine. Given such power over our landscape, what kind of a world do we want to live in? When we look beyond economic constraints, when we eliminate preconceived ideas of urban and ecological planning, when we free ourselves from the limits of materialistic thought, how do we envision the universe's age of fulfillment? In this section of the book, we draft a blueprint for reconstruction. We establish the design criteria for the earth of tomorrow and outline a plan to engineer a rural and urban landscape that meets our criteria.

That said, we must add a qualification, one anchored in the dynamics of the creative process. We cannot say exactly what the earth's urban and ecological infrastructure will be like decades or centuries from now, nor should we try to do so. The details await our invention, and we will learn as we go. But we can envision our future landscape in sweeping terms. We can create the framework of reconstruction. To do so, we begin by defining our limits and goals. What confines must we as builders work within, and are these limits real or imagined? What ideals of beauty, artistry, and functionality must we aspire to? Our task is to perfect life on earth, but what does perfection mean?

To establish the design criteria for rebuilding our planet's urban and ecological infrastructure, we begin by formulating the defining ideal to which our blueprint for reconstruction must aspire. When we take into consideration what we know about the earth, about our environment, and about ourselves as evolving beings in an evolving universe, what do we see as our blueprint's fundamental objective—as its guiding principle?

To achieve this end, we must accept what we have to this point devoted the book to establishing, humanity's status in evolution. As individuals and as a community, humanity is on the leading edge of the universe's advance. The cycles and thresholds of the universe's overall creative process unfold within us. We are that which embodies the capacity to envision tomorrow. We are that which incorporates the means to drive creation to its immediate and ultimate future. To bring into physical reality the earthly expression of the universe's age of fulfillment, we must look beyond the materialistic ideals of nineteenth and twentieth century science. Humankind is not a species like any other. We are not the product of the random turns of natural selection. We are not a mere organism caught up in the struggle to survive and the network of species interactions that we call the biosphere. The creative process builds on and creatively discards the past to create the future. Tomorrow unfolds through our thought and through our action. To rebuild the earth's urban and ecological infrastructure we, as individuals and as the human community, must accept our place in evolution—with all the humility and all the responsibility this embodies.

When we take as our own humankind's status in evolution, we have the means to put behind us an ancient idea, one we have discussed before, one born of humanity's move from a nomadic way of life to an urban way of life and one reinforced by the materialistic ideals of contemporary science. To achieve our goal of urban and ecological reconstruction, we must lay to rest the concept that man is separate from nature. From an evolution of consciousness standpoint, the notion that man must subdue nature and the notion that man is that which threatens nature have no meaning. In our internalization of the universe, the earth has no relevance without us. Other than through the reflective eye of humankind, the earth is an evolutionary relic, a sphere of organic matter that enwraps a sphere of inorganic matter. We and the earth exist as aspects of one becoming. We and the earth exist as dimensions of a single, unified thrust by evolution. Our task to perfect life on earth must go beyond ideas of stewardship and caring for the land. It must go beyond notions of survival, economics, and exploitation. It is our right and obligation to reshape the earth's urban and ecological infrastructure in support of the human endeavor. By virtue of our status in evolution, bestowed upon us by the passage of all prior time, evolution's trailing arrow is ours to command.

To be human is to be creative. We are driven to grow, to learn, to advance. We are compelled to create within ourselves higher states of consciousness and existence. We are motivated to perfect ourselves and our world. As beings in evolution, we embrace the human spirit to move forward, and any environment we create must be conducive to our individuality. It must provide us with the freedom we need to creatively express ourselves and to live as we see fit—to create and follow our personal road on evolution's journey. As important, our blueprint for reconstruction must serve as the medium for our creative expression. Not only must the world of tomorrow provide the environment we need to learn and grow, the act of molding that environment must contribute to that end.

Based on the preceding points, the guiding principle to which our design must aspire stands out. When humanity accepts its place in evolution—and we put behind us the notion that humankind is separate from the natural world and take as our own the right to command evolution's trailing arrow—our goal becomes to create a world that provides the freedom we as individuals need to creatively express ourselves. Furthermore, our objective becomes to embrace the design and building of this world as an avenue for the individual's creative expression.

> The objective of our blueprint for reconstruction is to establish the urban and ecological infrastructure that allows the individual to exercise his or her creative power and that through the process of planning and construction serves as an avenue for the individual's creative expression.

We want to shape a physical environment that allows every man and woman to evolve to his or her highest state of being and that—through the individual's growth and betterment, through our personal evolutionary journey—allows humanity to evolve to its highest state of being. And we want to evolve through the creation of that environment. The act of creation is as intrinsic to the task of earthly perfection as is perfection itself.

With the guiding principle for reconstruction in mind, we need to delve into and put behind us an issue that by now many readers have considered, the economics of rebuilding the earth's urban and ecological infrastructure. As we establish our design criteria, what are the limits to reconstruction?

Prevalent in present-day intellectual circles is an ideological value that since the 1960s has been driven into the mindset. It is the notion that economic growth is intrinsically bad. Instilled at our colleges and universities, and driven by the socialist and environmental movements, it is the view that economic prosperity, and in the larger sense development and industrial and technological progress, is in some way wrong—that humanity prospers at the expense of the environment, that the United States and other developed nations prosper at the expense of the third world.

Case in illustration, the 2009 Copenhagen Agreement to limit "greenhouse" gas emissions and, in the minds of advocates of anthropogenic climate change, prevent "catastrophic" global warming.[1] A key issue during the talks that led up to President Barack Obama's signing of the agreement was the need to limit greenhouse gas emissions from undeveloped nations. During the course of negotiations, United States Secretary of State Hillary Clinton committed her country to pay third world nations a sum of 100 billion dollars a year to limit greenhouse gas emissions. What does this mean? The question of where the United States—at the time of the agreement strapped with debt and printing vast sums of currency to devalue the dollar—would get the money aside, Clinton has committed her nation to pay the governments of undeveloped countries to do the one thing that, in those nations, will limit greenhouse gas emissions— slow development, restrict modernization. The view that development and human progress is undesirable is predicated on the twelve-thousand-year-old notion of scarcity. Based on the assumption that resources are inherently limited and human material wants are inherently unlimited, the political elite in the United States has, with the signing of the Copenhagen Agreement, made the commitment to lock families in undeveloped nations into poverty and the often harsh life of subsistence agriculture.

As we established in *Threshold to Meaning: Book 2, Economics of fulfillment*,[2] the notion of scarcity is only relevant with respect to the human demand for resources. In terms of material resources, what would it take to develop the undeveloped world: to build roads, schools, and homes and to reap the benefits of modern agricultural practices? It would take energy, steel, portland cement, sand and gravel, and some copper and other minerals.[3] What would it take to rebuild the earth's urban and ecological infrastructure and, at least from the standpoint of our physical

---

1. See chapter 10, *Climate*.

2. See *Book 2, Economics of Fulfillment,* chapter 5.

3. See chapter 9, *Energy*.

surroundings, perfect life on earth: energy, steel, portland cement, sand and gravel, and some copper and other minerals. The shortages of resources we deal with in today's world are not the product of limited natural supplies. They are a product of scarcity-based economics and an obsolete interpretation of the "natural" world. It makes no sense that evolution would bring humanity to the point of realizing its purpose in existence without making available the material resources to further that purpose. For reasons grounded in the evolution of consciousness view, there is no limit to the material resources we need to perfect life on earth.

Neither do we face a limit on the availability of human resources.

When we think of a construction project, we typically see it in terms of codes and inspections, in terms of unions and labor laws, in terms of architects and engineers who oversee a general contractor, who oversees what may be a long list of subcontractors. Construction is a legal, political process, inseparable from government and the demands of interest groups—in nearly every respect administered from the top down. That said, to be successful, a construction project must be well planned and systematically implemented. A project must be run by an individual or by a group of individuals who know what the project will look like when completed and what materials and building procedures are needed to bring it from conception to physical reality. There must be an organizational structure, but in the future this structure will not be dominated by the top-down, political process that defines construction today.

To understand this, we must draw on our idea of collectivity. As we have spoken about in earlier chapters and developed at length in the earlier books, human social structure has evolved from more collective to less collective forms. This decrease in collectivity has been driven by an increase in the autonomy and consciousness of the individual, which is rooted in the dynamics of evolution. The greater the autonomy and consciousness, or substance of character, of the individual, the stronger and more intimate the social bonds the individual can form and the more nestled and interwoven the social structure that results. Correspondingly, the less the social structure is sustained by some form of government or ruling elite. We can maintain order in society in two ways: by imposing it from the top down, through government and the police power of the state, or by allowing it to arise from the bottom up, from the individual. As, over time, social structure became less collective, the individual became more important and rule by a governmental class became less important.[4] A person living in Ancient Egypt saw his or her activities under the harsh management of a ruling elite, the leaders of which often had the status of gods. A person living today has many educational and other opportunities and far greater freedom to choose his or her direction in life. As time passed, we increasingly achieved social order because we chose to do so; not because we were told to do so. The greater our substance of character as individuals, the more cooperative the society that results.

The rebuilding of our planet's urban and ecological infrastructure is not something that a ruling elite or administrative class will conceive and administer. To realize our dream of planet-wide engineering, we must form organizational structures. There must be individuals or groups of individuals who envision a project and oversee its completion. In a way that contrasts with the building practices of today, however, the organizational structures we create in the future will arise from the bottom up. When we grasp the direction of evolution, when we know where the universe is headed, we take as our own the goal to bring forth its destination. Grounded in the enlightenment of the individual, we will form the organizational structures we need to facilitate our personal drive to create fulfillment.[5] The realization of our individual passion, in turn, will propel the universe and the human experience to higher levels of achievement. The greater our substance of character as individuals, the more cooperative our joint ventures and the more grand and ambitious the projects we can achieve—embracing the earth in its entirety.

As such, the concept of labor as we traditionally think of it has no meaning. The labor to rebuild our planet's urban and ecological infrastructure is not something that a ruling elite must hire or conscript. The labor for reconstruction is all of us, doing what by our nature as creative beings in a creative universe, we are driven to do. When we are doing what we love, work is not work. It is our passion, our purpose in life. Reconstruction of the earth's urban and ecological infrastructure is the medium through which we express the human drive to build. Humanity is what reconstruction is all about, and we can do so to whatever extent is necessary to realize our creative potential. There is no limit to the human resources we need to perfect life on earth.

4. See *Book 1, Evolution of Consciousness*, chapters 11, 12, and 14, and *Book 2, Economics of Fulfillment*, chapters 1 and 3.

5. See *Book 2, Economics of Fulfillment*, chapter 10.

There may be no limit to the material and human resources available to rebuild the earth's urban and ecological infrastructure; but, as by now the reader must have asked, how will we pay for it? How can humanity afford to reshape the world's cities and countryside on the scale of the planet in its entirety?

At its core, an economic system—socialistic, capitalistic, hunting and gathering, economics of fulfillment, or any other—rests on the creative energy of the individual. Every economy exists as a result of our work, our dreams, our ambition, our skill, our drive, our knowledge, our desire to do something better or different. For our creative drive to manifest in physical reality, we need material and human resources: energy, raw materials, partners, and colleagues. Our economic philosophy—and the systems through which we implement that philosophy—is the tool that allows us to direct our use of the material and human resources at our disposal to attain a desired end.

Under socialism-capitalism, the predominant method through which we achieve this objective is to assign the elements of construction a monetary value. We create labor and material budgets and track our expenses. In college, an engineering student studies the strength of materials; he or she also studies accounting and job costing. Contractors fight against time to bring their jobs in on or under budget. How often has a government energy, defense, infrastructure, or other project cost the taxpayers as little as they were told it would cost? The choice to assign a monetary value to our steel, concrete, hours of human activity, and other construction elements, however, is not grounded in nature. It is a function of scarcity-based economics. In a scarcity-based economic environment, the monetary valuation of construction elements enables us to allocate what, by virtue of our economic system of beliefs, we take to be scarce resources. Under socialism-capitalism, coherent construction on a planet-wide scale would not be possible. Socialism-capitalism is not the economic tool that evolved to allow for human activity of such a scope. Perfection of life on earth would be inconceivable. It would cost too much.

The economic tool of earthly perfection is the economics of fulfillment ideology, and the systems we devise to implement it. Economics of fulfillment, and humanity's perfection of life on earth, are aspects of a single movement in evolution, dimensions of humanity's transition from the universe's evolutionary period of understanding to the universe's evolutionary period of fulfillment. Under economics of fulfillment, money has no intrinsic value. We need not assign the elements of construction a cost. In our quest to perfect life on earth, we will implement whatever analytical tools—whatever schedules, whatever labor and material lists and budgets—we need to channel our resources in the best possible way, but monetary cost is irrelevant. Monetary valuation has no meaning.

Economics of fulfillment allows us to move beyond the utilitarian architecture and engineering of the nineteenth and twentieth centuries and embrace a greater ideal of perfection. We can afford to rebuild the earth's urban and ecological infrastructure for the simple reason that in the universe's age of fulfillment the concept of "afford" as we today define it has no meaning. Humanity's reconstruction of the earth's urban and ecological infrastructure is not about costs, budgets, and the allocation of artificially scarce resources; our goal is to advance who we are, to further humanity on its evolutionary journey.

We face no limit to the material resources we need to engineer the earth's urban and ecological infrastructure. We face no limit to the human resources we need for reconstruction, and our economics of fulfillment ideology allows us to bring our material and human resources into play without the restrictions imposed by cost and money. But we do face a limit, one that exists within every man and woman, one that each of us must rise above. To undergo the task of earthly perfection, we must overcome the barriers we place on our freedom of thought. Within the realm of common sense—no elevators to the moon, no bridges across the Atlantic, no science fiction teleporters in lieu of cars and transit systems—the earth is ours to shape. Urban and ecological rebirth is limited only by our dreams, only by our inventiveness, only by our willingness to extend the reaches of what it means to be human.

That said, it is time to look past the philosophy of reconstruction and, in our task to establish the design criteria for humanity's perfection of life on earth, delve into the details of construction. The guiding principle of our blueprint for reconstruction of the earth's urban and ecological infrastructure is broad and encompassing. We have the material and human resources we need to do the job; and, as economics of fulfillment takes hold, we will have the economic framework we need to rise above the artificial barriers of monetary costs and budgets. On this foundation, we can build any world we imagine. In this world, what do we want our countryside to look like? What do we want our cities to look like? How do we want to get around and to communicate? Here we step onto familiar ground, one the reader may find reassuring. For reasons that are different than our own, many of the goals we put forth have been long valued by planners and environmentalists.

It goes without saying that we want wilderness areas. The preservation of certain traditional ecological systems is of unquestionable value. Long ago, Muir, Marsh, Thoreau, and others brought into consciousness the spiritual value provided by ancient ecological systems. These founders of the environmental movement were insightful. Wilderness gives us a sense of belonging in the universe. Wilderness allows us to look back in time and to come to terms with the evolutionary path that brought us across the threshold of reflection, through the universe's age of understanding, and into the universe's age of fulfillment. The simplification of the biosphere began long before humankind played a direct role; but, as those in command of evolution's trailing arrow, we make the design choice to, in certain parts of the world, stop or even roll back organic decomplexification. Humanity needs wilderness.

As the engineers of evolution's trailing arrow, we also make a design choice in the management of the earth's intermediate zones, those regions that are not wilderness but that are not as heavily developed as arable land and areas closer to urban centers. Here, our goal is not to preserve traditional ecological systems but to manage traditional systems in a way that meets specific human needs. These needs may be recreational: skiing, hunting, fishing, boating, and backpacking. They may be practical. As we will see, our need for water, mineral, and timber resources in the future will not be as great as it is today. But we will still need raw materials. Our goal in the management of the earth's intermediate regions is not to turn back the clock and return these areas to "pristine wilderness" but to stabilize ecosystems at a level of complexity that allows us to interact with them without disrupting their equilibrium, or range of equilibrium states, to such an extent that we drive them across the threshold to a less complex, and thus to a less useable, ecological level. We want to manage intermediate zones in a way that allows for our use over time.

The earth's agricultural lands require a more focused approach. An agricultural system is not a natural ecological system but an engineered ecological system. It is a set of human determined ecological interactions maintained through the input of labor and materials. In this process, we must move beyond the idea of "sustainability" as professed in environmental circles and university lecture halls. To produce the output of the agricultural products we need and desire, we must incorporate into the system the necessary water, fertilizer, and other inputs. We can achieve this with manures and the use of other "natural" fertilizers only to a point. There is not enough livestock in the world, and enough manure produced, to support agriculture on today's scale without the use of chemical fertilizers. We can and will always improve our agricultural practices, and advances may incorporate greater use of "natural" inputs, but to limit the use of chemical fertilizers would be to dramatically decrease the world's food production, create an artificial scarcity of goods. Agriculture in the future must be based on our best understanding of chemistry and not on stagnant environmental ideology. The idea that modern agricultural practices are not sustainable is unfounded. The "green" revolution is nearly a century old. In the past, we have made mistakes in the use of certain agricultural practices, and we have learned from our mistakes, but overwhelmingly agricultural land today is as healthy and productive as at any time in its existence.[6] The green revolution is sustainable.

We need agriculture for food, fiber, and other organically based products. We also need agriculture for another reason, and it is no less important. Today, in technologically developed regions of the world, agriculture is a scarcity-based economic process—land, labor, and capital. Yet, farming and the connection to the land that farming provides can be a remarkably fulfilling experience. Our design objective is to engineer our agricultural practices in a way that not only meets our need for organic matter but that fulfills the creative needs of the farmer. Our design criteria calls for a landscape where the agricultural process is driven by the creative desires of those who work the land. We want to embrace the tools of science and reestablish the family farm—farming as a way of life.

As for our urban environment, our goal is two-fold: First, we must engineer our cities to allow for the freedom, mobility, and above all the creative inspiration of the individual. Second, we must engineer our cities to allow for the transformation of social structure that the freedom, mobility, and creative inspiration of the individual will bring about.

6. The commonly held belief that modern agricultural practices have led to excessive erosion, depleted top soil, and a decrease in soil fertility is not supported by observation. Over-farming is an issue in undeveloped nations where fertilizers are not available, and certain agricultural practices, such as the dry-land farming techniques that led to the dust bowl of the 1930s in the United States, have had environmental consequences. Properly implemented, modern agricultural practices have led to deepening top soil and healthy and sustainable agricultural environments.

As we have discussed to the point of belaboring the idea, and for reasons covered at length in the earlier books, human societies have grown less collective over time. We have formed social arrangements that are less communal and more nestled, intricate, and strongly bonded. The city of tomorrow cannot be rigid or in some extreme way mechanistic. In contrast to the land-use planning ideals of today, it cannot be based on the notion of "for the collective good" imposed by a ruling elite through law, regulation, and central control. Our urban design must catalyze and allow for our personal growth and for the manifestation of that growth into less collective forms of social organization.

To accomplish this, our cities must have a structure where smaller urban units are arranged to create successively larger urban units. The basic urban unit, we will call it the *urban community*, must support a flexible number of more intimate levels of social order down to the extended and immediate family. It must also be the building block of a coherent city social structure and, as such, the building block of a coherent regional, national, and ultimately global social structure. The disperse geometry of today's city reflects a high degree of collectivization; its engineering mirrors contemporary social order. To accommodate the social structures of tomorrow, the city must not be based on geography but on centers of human activity. It must have a nestled and interwoven design.

In this urban framework, we must also take into account the practical aspects of urban life. We must engineer communication and transportation systems that allow us to freely interact with one another. We must devise ways to run utility lines without stringing wires on poles or burying pipes under streets. We must devise ways to efficiently transport goods and to process and dispose of waste. Today, we structure our lives around our urban environment—around money, safety, and traffic. In the future, we will structure our urban environment around our lives. The human being, and our nature as evolving beings in an evolving universe, will be the incontrovertible center of urban planning.

The guiding principle of our plan to reengineer the earth's urban and ecological infrastructure is to create an environment conducive to human freedom and creativity and to provide for our creative expression in the building of that environment. To adhere to this principle, we must preserve and develop our rural lands and engineer our cities to provide for mobility, communication, and social decollectivization. At the heart of every design choice must be the individual. Our task must be to provide every man, woman, and child with the freedom and material resources to follow his or her path in evolution. Some people prefer a rural lifestyle, with the solitude and introspection this provides. Others prefer an urban lifestyle, with the sense of community and human energy this provides. As architects and engineers, it is not our role to judge another person's evolutionary road. It is not our place to dictate where and how others should live. Neither, in an economics of fulfillment world, is it our job to allocate resources and limit opportunity and avenues for personal growth. The earth's urban and ecological landscape is the canvas of all. Our task is artistic and functional. It is the perfection of life on earth and the building of that which allows for earthly perfection.

6

The Countryside

IN THE LAST CHAPTER, we established the criteria our blueprint for re-construction of the earth's urban and ecological infrastructure must address. In this chapter, we draw up a plan to rebuild our rural landscape in a way that supports our criteria. Our design will include wilderness areas, intermediate zones, and agricultural lands. It will address standards of rural construction and the development of rural highway systems and rural communities.

We begin with the first of these elements, wilderness. Our plan must include traditional ecological systems, but this means more than cordoning off a stretch of woods and hugging a tree.

Today, we think of wilderness as a single class of land. In the future, we will create two categories of wilderness. In the first will be those few remaining large tracts of land: parts of Alaska, Siberia, the Canadian Northwest territories, and other remote regions of the world. Our goal in these areas is to stabilize traditional ecosystems. We want to stop the clock and roll it back. We want ecosystems as they were one hundred, two hundred, three hundred years ago. Our aim is to minimize human impact and to stall or reverse evolution's trailing arrow.

Access to the world's most remote wilderness areas, however, is difficult. In part, this is because these areas are hard to get to. In part, it is because on entering them we must take into account their wild nature. If we were to setout into a wilderness as it was in the 1800s, we would have to do so as we did in the 1800s. Take the Alaskan bush and the Canadian Yukon and Northwest Territories. In these areas, we find polar bears in the north and brown and Kodiak bears in the south. Only the most daring or ignorant stray from tourist lodges and trails unarmed and without at least some knowledge of predator behavior. Similarly, in Tanzania's Serengeti National Park, visitors are usually escorted by trained and armed park personnel, and tours are often conducted by vehicle. To access our planet's truly wild places, we must accept them for what they are—wilderness. This may mean we hike with a dog, a guide, and a rifle.

Given the outdoor skill of the average wilderness lover, and a trendy aversion to firearms, the world's most remote places are not within reach of most of us. We also need wilderness areas that are tame, wild places where grizzlies and other animals are not a threat, where trails are mapped, marked, patrolled, and maintained. This description fits many wilderness regions in Europe and the contiguous United States. Areas are within reach of a road, and a hiker with a minimum of skill and equipment can make peace with the earth without risk to life and limb.

With regard to our second category of wilderness, we must not only manage areas based on an understanding of natural systems but also of human needs. Is it appropriate to reintroduce the grizzly into wilderness areas in Oregon and Washington, an act that as we mentioned would effectively close these areas to the casual hiker and backpacker and increase impact on tamer wilderness lands? For more than a century, residents of the American West coexisted with the cougar, or mountain lion. In recent years, animal rights groups have succeeded in limiting cougar hunting, in particular with the use of dogs. The cougar population has exploded, and animals too young to remember hounds and hunters no longer fear men and dogs. Cats maul hikers, decimate elk and deer populations, prowl the suburbs for pets, and are driving big horned sheep in California into extinction. For millions of years, ecosystems have realigned, and species habitats have come and gone. The cougar and grizzly may have once roamed throughout the West, but they have no "God given" right to their traditional habitat. As managers of evolution's trailing arrow, the place of a species in the world is ours to determine, and we must do so with the species and with ourselves in mind. As engineers of evolution on the universe's trailing edge, we provide for wilderness lands not for the sake of their existence and innate value in "nature" but to further the human endeavor on earth.

We must also design our planet's intermediate zones with deliberation. We must provide for stable ecological systems, recreation, and resource extraction.

To illustrate an intermediate zone design, we will expand on last section's fire control example. Intermediate zones in the Western United States are often forested and dotted with homes, towns, lodges, ski resorts, and mining and logging facilities. Because of human activities in our intermediate zone lands, foresters cannot let nature take care of itself, as the environmentalists would like to see. The threat to life and property is such that it is not acceptable to let fires burn where they may. We cannot and will not depopulate the rural West and let fires sweep across the landscape as they did a thousand years ago. Foresters must introduce fire into the forest ecosystem in a managed way, through what is called a *prescribed burn.*

In a prescribed burn, brush is cut, fire-lines are built, the forest is ignited, and the burn is controlled. To cut brush, prescribe burn, or manage intermediate zones in anyway, we must have access. We must have roads. Roads scratched into a hillside to harvest timber or to fight a fire can be damaging. Established and maintained roads are another matter. A road engineered with topography and soil conditions in mind will have minimal environmental impact. Rather than threaten the interaction between species, a properly built road becomes part of the ecosystem.

Imagine an intermediate zone forest in the Western United States that is accessed through an extensive network of thoughtfully placed, well-engineered, and regularly maintained roads. Unlike the haphazard and poorly built and maintained backcountry roads of today, we would not construct this road system for short-term use. It would provide permanent access to the forest. In effect, our system of maintained roads would partition the forest into carefully determined management units. Users would have access to recreational sites and wilderness trailheads. Foresters would have access to timber resources and be able to reduce fuel loads and prescribe burn with little impact. Our road system would serve as a network of permanent firebreaks.

Our road system would also enable us to harvest resources with minimal impact. Rather than lay out a timber sale and build or reopen roads to access it, we would have ongoing access to timber resources. In recent years, the annual timber allocation in the state of Oregon has hovered around three billion board feet, with less than one billion board feet harvested from federal lands. Though low by historical standards, when taken from a series of small timber sales, this amount could impact an area of forest in a destabilizing way. When spread over the state's entire forested region, however, the harvesting of three billion board feet, or considerably more, would have almost no impact. Our network of permanent roads distributes use over a large area. It allows us to draw resources where it is appropriate, a tree here and there, and at a stage of their development when it is most advantageous. To a greater extent than today, timber harvest becomes a tool to improve forest health and diversity.

In this regard, our network of roads also allows us to clear-cut a patch of forest here and there. The scalping of thousands of acres of Western forest land in the 1950s and 1960s gave clear-cutting a bad name and led to a near abandonment of the practice on state and federal land and strict controls on private land. Open spaces in the forest are a natural outcome of fire, and certain animals cannot survive without them. Decades ago, vast elk herds roamed the forests of Northern and Central Idaho. Fire control and a reduction of clear-cutting led to a sharp decline in population. Adults need open spaces to graze, and calves bedded down in dense forest have no warning against bear and mountain lion able to steal up on them through the undergrowth. As the hunter will tell you, we do not often find elk in untouched and overgrown wilderness land. We find the animal in areas that have been recently logged or burned.

Similarly, the Northern Spotted Owl nests in "old-growth" forests in the United States Pacific Northwest. To save the "endangered" spotted owl, environmentalists in the 1990s lobbied to ban logging in these forests and politicians eager for votes and campaign contributions from the "green" community stopped almost all timber harvest on public lands. Today—after laws were passed, logging curtailed, an industry devastated, and thousands of family-wage jobs lost—the northern spotted owl is threatened by the common bard owl, better adapted to the dense, overgrown forests our good but misguided intentions created.

When managing our intermediate zones, our goal is not to return ecosystems to an earlier evolutionary state. It is not to turn them into wilderness or to save every tree or blade of grass. Our goal is to stabilize ecosystems at a level of complexity that sustains our use—that integrates plant and animal species with human activity. Through fuel reduction, prescribed burns, and timber harvesting, we manage forest health and diversity. Through hunting and fishing, we manage wildlife populations. In some areas, we will have mining, ski resorts, lakeside lodges, and other concentrated forms of development. In other areas, we will have activities with low impact: hiking and sightseeing. But to manage ecosystems in the earth's intermediate zones, we must be active at every level. For millennia, Native Americans set Western forests ablaze to clear brush and create the open, diverse, wildlife friendly forests that the early Europeans found and took to be natural. We cannot separate ourselves from the management equation. Stabilized to meet our needs, intermediate zone ecosystems cannot take care of themselves.

Management is even more critical on agricultural lands. To maximize output and to meet the creative needs of the farmer, we must break farmland into chunks that a farmer, a farming family, or a farming social unit can take care of. In today's economic climate, the trend has been toward consolidation. Large corporate concerns have bought out family farms to the point where, in many parts of the world, agriculture has become a centrally controlled production process. As we have seen in the most extreme form of agricultural consolidation, the collective farms of the former Soviet Union, which were so unproductive that a disproportionate share of the nation's food was produced on tiny plots the government allowed workers to cultivate for themselves, this structure does not lead to maximum production or to a satisfying relationship between the land and those who work the land. Driven to turn a profit, corporate farms are vastly more productive than collective farms, but they do not nurture a personal relationship between the land and the farmer and the sustainable production this yields.

Our plan for agricultural reconstruction calls for farms sized to accommodate the crop and the farmer. The growing of wheat or alfalfa will take place on large tracts of land. The management of these crops is heavily mechanized, and a farmer can work big areas. Grapes and tomatoes require more attention and will be grown on smaller, more intimately managed plots. One farmer or one farming family, however, may not be able to grow certain crops at a practical scale. In the economics of fulfillment world of tomorrow, social units of farmers, most reasonably in the form of a business or company, will emerge as needed to provide the organizational framework to coordinate farming activities and to interact with the land in the most nurturing and fulfilling way. As in the overall management of the earth's urban and ecological infrastructure, the autonomy and substance of character of the farmer will allow for the creation of the social structures necessary for the farmer to fulfill his or her creative needs.

Our approach to agriculture also allows for better integration of arable land into intermediate zone land. In the state of Idaho, cattle ranching is a traditional way of life. To fatten his cattle, the rancher grazes his herd over a large area. Many find ranching a rewarding way to earn a living; and, with disease and other problems in feedlot production, it may become a more important method of meat production in the future. Cattle grazing also fills a void left by the decline of native grazing species, in particular the bison. Under economics of fulfillment, the farmer is free from the need to show a profit and is thus better able to manage his or her herds to improve their quality, the quality of the rangeland, and the quality of his or her life. The management of appropriately sized tracts of land also allows for better use of pesticides and fertilizers and for better use of water, a critical concern in many parts of the world.

Future agricultural practices will reflect a growing understanding of artificial ecosystems and a better integration of artificial ecosystems into natural ecosystems. By establishing the conditions where the farmer meets his creative needs through an intimate relationship with the land, we establish the conditions where the outcome of his creativity can unfold. Agricultural land will be cared for to an extent that matches or exceeds the best farming operations of today, and we will realize its potential for production and perpetual yields.

Like today's countryside, the countryside of the future will also be populated. Farmers may choose to live on their land. Others may simply prefer a rural lifestyle. I grew up in the rural West and know firsthand the appeal of the countryside with its rivers, lakes, horses, wildlife, and access to ski resorts and wilderness areas. It is not a lifestyle all or perhaps most will choose in the future. Not everyone takes enjoyment in isolation, in fixing a fence in a rainstorm, or in chaining a truck in a blizzard. But, as today, there will be some who seek this life, some who choose to make the countryside their home. In the future, though, we will interact with the countryside in a less haphazard way. Above all, we will build differently.

Whether a home, lodge, or other building, our design must be compatible with its environment. It must be aesthetically pleasing. It must also be integrated into its environment in a practical way. We must engineer our structure to withstand the rigors imposed by the ecosystem in which it is built. Homes must be able to withstand floods, tornadoes, wildfires, hurricanes, earthquakes, and any other hazard nature throws their way.

To meet this design objective, we begin with a little common sense. Not even the strongest home can withstand a severe flood. The solution is as simple as not building on a floodplain. Many areas, however, experience occasional and mild floods, and it may not be practical to avoid building in these areas altogether. The answer is to build the structure to withstand the area's floods. Living space may be elevated. Basements may be waterproofed.

Given this and other natural hazards, the building material of choice will not be wood, unreinforced stone or masonry, or any other traditional home construction material. Not even the best-built wood structure can withstand wind, water, or wildfire. Wood is proclaimed to be a renewable resource, but this is only true if a wooden structure has a lifespan equal to the time it takes for a tree to mature. How long does it take for a forest to reach old growth status, one hundred years, two hundred years, three hundred years? How long does the average wood structure last, thirty years, forty years, fifty years? Throughout much of the world, timber is not available. Masonry in the form of stone blocks, sun dried mud bricks, kiln fired ceramic bricks, and cast concrete blocks is the standard building material. Unreinforced masonry structures do well in fire and moderate wind but perform dismally in earthquakes. The tremendous loss of life experienced when earthquakes strike undeveloped parts of the world is almost entirely due to the collapse of unreinforced masonry buildings.

The material of choice is reinforced concrete. Modern concrete is made by mixing sand and gravel with *portland cement*, a mixture of heated and pulverized clay and limestone. In the 1800s, we discovered that if we embedded steel mesh or rods in the concrete the resulting material would have concrete's compressive strength and steel's tensile strength. Today, reinforced concrete is a standard building material for large structures. In the future, it will be a standard building material for small structures. *Figure 6* shows a perspective view taken from the working drawings of a large country home of the future.

Fig. 6. Country Home. The illustration shows the front perspective taken off the working drawings for a 3,500 square foot country home of the future. Note the classical European features and proportions.

Architecturally, the design uses principles of symmetry, repetition, and Fibonacci proportions. Structurally, the design is *monolithic*, built by casting concrete between insulation and brick walls to create a reinforced box structure, *Figure 7*.

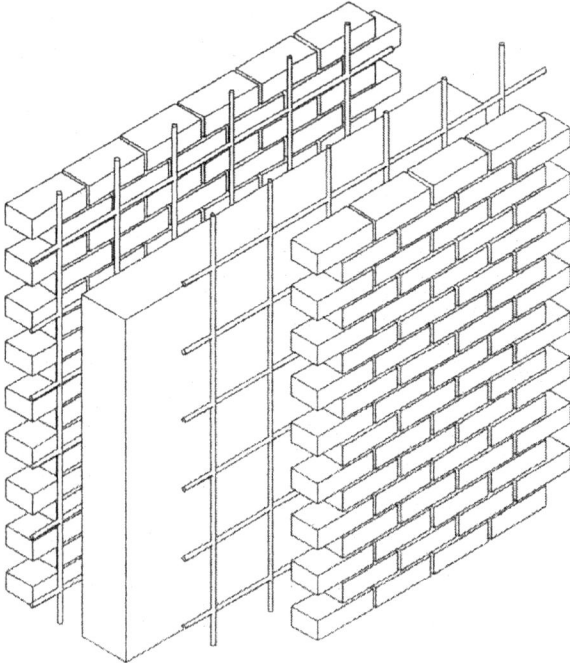

Fig. 7. Country Home Construction. The country home shown in Figure 6 is constructed by casting concrete between insulation and brick walls to create a monolithic box-structure. In this design, the total wall thickness is 23-5/8 inches, with reinforcement placed on an 8 inch center. The use of ceramic bricks anchored to the concrete reduces the need for molds, enhances the home's thermal behavior, and produces pleasing interior and exterior wall surfaces.

This structure's strength and the anchoring of bricks and roof tiles enables the home to withstand extreme earth movement and, like a concrete tornado shelter, debris hurtled by winds in excess of 300 miles per hour. Fire is not an issue, and insulation, concrete's high specific heat, and hot and cold temperature reservoirs built into the home's foundation and ventilation space turn the entire structure into a passive solar radiator and collector, minimizing the need for heating and cooling. The absence of paint, wall cavities, and structural wood eliminates pests, toxic mold, maintenance, and water damage.

Architecture is subjective. Environmentally responsible construction is not. Those who pioneered North America built log cabins because they could throw up a cabin in two weeks and beat the winter snows. Today, we see a log cabin in the woods as romantic. From the standpoint of the forest ecosystem, it is food for carpenter ants and tinder-dry fuel for a wildfire. For buildings to support and positively interact with their environment, they must withstand the rigors imposed by their environment.

The earth's future countryside will be dotted with farms, homes, and other structures. It will also be dotted with rural communities and be overlaid by a transportation system that links rural communities and ties together the countryside.

It is popular to look down on the automobile, and we spoke about it critically in an earlier chapter. Yet, the automobile has advantages that will secure its place in the future countryside. In relatively unpopulated areas, there is no better way to get around. How else can we travel our network of intermediate zone roads and bring in the manpower and equipment to prescribe burn a forest? In what other way than by car can we get to a wilderness trailhead? The subway? The automobile is the heart of a disperse transportation system, and this type of transportation system works well in an area with a small and widely distributed population. We want to be able to get from the vicinity of any geographical location to the vicinity of any other geographical location. We will address issues of energy, pollution, and fuel economy in a later chapter.[1] For now, it is enough to realize that in the countryside the car and in particular the truck are here to stay.

The highway systems we drive on, however, will be different than those of today. Contemporary highway design is driven by environmental and other political concerns and by budgets and construction costs. As a result, our roads are not particularly safe, well designed, or adequately maintained. In the future, transportation itself will be the planner's overriding design objective. We will close old highways or tear them apart and rebuild them to new standards. We will lay out new highways in the most coherent way. We will use pavements formulated to last. We will design

1. See chapter 9, *Energy*.

lanes to carry traffic loads where they need to go. We will isolate bike and pedestrian traffic from automobile traffic. As anyone who has lived in the country knows, the power goes out in almost every thunderstorm. We will not bury utilities under roads or string them on poles next to roads but in ways where they will function reliably and where we can access and maintain them. In every respect, we will design the highway system of the future with the needs of the rural resident in our mind—safe, coherent, reliable, functional, the best we can envision, the best we can engineer and build.

The highway system of the future will also link farms and homes to the *rural community*. In many ways, the rural communities of tomorrow will be like the best countryside towns and villages of today. Communities will have parks, plazas, streets, and buildings. We will integrate commercial spaces with stores, homes, schools, and restaurants. Buildings, however, will be engineered to higher construction standards, at least as rigorous as those we adopt in home construction. We will access rural communities by car; but, in the community, primary transportation will be on foot. For this reason, the rural community will be small by today's standards. If the community were too spread out, we would have to drive and to deal with the problems of the automobile. To be practical, any large rural community would have to be made up of coherently linked smaller communities. By virtue of their unity and integrated design, the rural community of tomorrow promises the best of small-town life in the countryside.

Thus far, our blueprint for reconstruction calls for true wilderness and for wilderness managed for safety and access, able to accommodate the individual who has minimal equipment and survival experience. It calls for intermediate zones where we stabilize ecological systems to provide for our use and exploitation. Our design includes farms, homes, rural communities, and a highway system that ties the countryside into a cohesive unit. As earlier, the examples I cited drew on my experience in Western North America. I live and work in this region and know its wild places well. Parallels can be made to ecological systems around the world. The details will be different in Asia, Africa, or South America, but the principles will be the same. We will have wilderness areas and intermediate zones. We will live, work, and farm in the countryside.

7

The City

Tomorrow's rural communities will be a remarkable place to build and to go about our lives. The rural community will also link the countryside to areas of the earth that will have almost nothing in common with any that exist today. In this chapter, we draw a plan to reconstruct our urban landscape. Our design will include the urban community, the city unit, and the city. It will address construction, communication, and transportation.

The focal point of the city will be what in chapter 5, *Design Criteria*, we called the urban community. We defined the urban community as a social unit cohesive enough to support nestled social structures down to the level of the extended and immediate families. We also defined it as a social unit cohesive enough to serve as the building block on which we would construct the larger social structure of the city. What will be the urban community's dimensions, and how will we tie urban communities together to create a larger urban framework?

The task of designing our urban community may at first seem intangible. But, by looking at the urban community in the context of human social and physical needs, its form stands out. To establish the layout of the urban community, we begin with a simple fact. A disperse transportation system does not work in a densely populated area. As population increases it becomes difficult and ultimately impossible to build the roads, parking lots, support facilities, and everything else we need to get around by car. Cars, trucks, and highways work in the countryside where the population is low and spread out, but they have no place in the city. Most of us equate the automobile with freedom. If the automobile increases our mobility, it increases our freedom. If the automobile limits our mobility, it limits our freedom. Any attempt to impose the automobile on the city of the future will result in nothing more than the urban mess we live in today. Limiting use of the automobile for reasons of environmental ideology is not the issue. We need an urban design that provides greater mobility and greater freedom.

As *Figure 8* shows, imagine a circular plot of land with an area of one square kilometer, or 250 acres. This would give it a diameter of about 1.13 kilometers, or somewhat less than a mile.

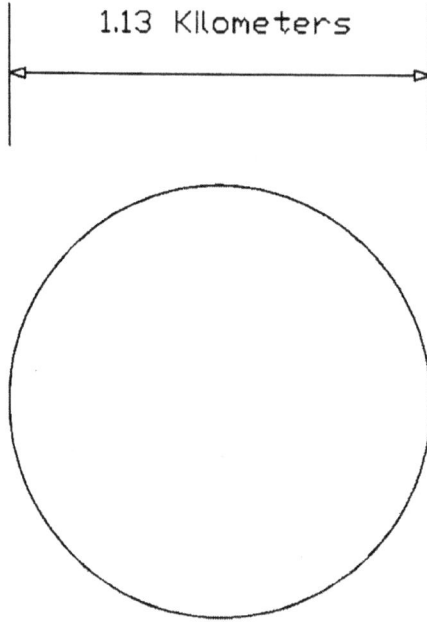

Fig. 8. Urban Community. The one square kilometer circle of the urban community would place everything in the community within a comfortable twelve minute walk.

This shape and size is significant for two reasons. First, it is large enough to, if properly utilized, house an urban community with a population sufficient to support a diverse economy and to embody a social structure with a large number of nestled levels. Second, it is small enough so that every place in the urban community would be assessable by foot. The average person walks at a pace of about six kilometers, or a little less than four miles, per hour. This puts every place in our one square kilometer community within a twelve-minute walk. In a densely populated area, walking is the only practical way to get around. Even today, we must at some point get out of our car and walk. There is no alternative to walking.

Too many people on bicycles can be nearly as troublesome to plan for as too many people in cars. The basic component of our urban transportation system must be travel on foot, and we must engineer our urban community to meet this requirement.

Now, as *Figure 9* shows, imagine that the one square kilometer circle of our urban community was built relatively close to the one square kilometer circles of other urban communities and that each urban community was linked by some form of transit system.

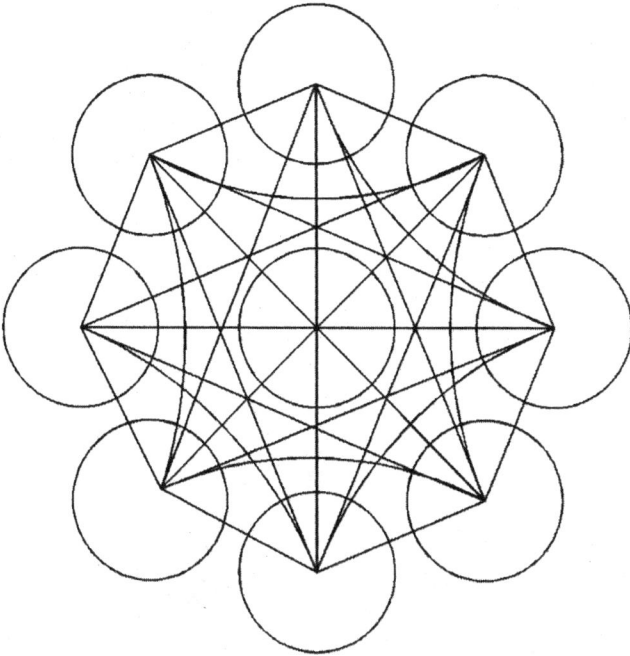

Fig. 9. City Unit. The illustration shows a city unit composed of nine urban communities. The lines represent transit links between communities. Note that each urban community is only one transit stop away from any other urban community.

We have created a larger urban division, a *city unit*. Because each urban community would have a population sufficient to maintain a diverse economy and a nestled social structure most of our activities would take place within the urban community. We could think of the urban community as our hometown. When, however, we needed to travel to another community, we would catch the transit. To attend a class in a neighboring community, a student would walk to her community's terminal, catch the transit to the community where her class is held, and walk to her class.

In contrast to contemporary transit systems, this scheme works for two reasons: First, we are not trying to transport everyone in the city at the same time; most of what we need to do we can take care of in the urban community. Unlike rush hour traffic today, where it seems everyone in the city is trying to get anywhere but where they were at, the bulk of our movement would be in the urban community and thus be on foot. Second, our transit system links population centers and not geographical locations. We travel between centers of human activity and not between blocks, our bus or train stopping every few minutes to let someone on or off.

If we locate the transit terminals in our urban communities in the center of their one square kilometer circles, it would be roughly a six minute walk to any terminal. If the transit ride took three minutes—a two kilometer distance at a speed of forty kilometers per hour—no place in the city unit would be more than a twelve minute walk and a three minute transit hop away, fifteen minutes.

Now, as *Figure 10* shows, imagine a larger urban structure, one made up of city units, each linked by a transit system. This configuration creates the more complex urban body of the *city*.

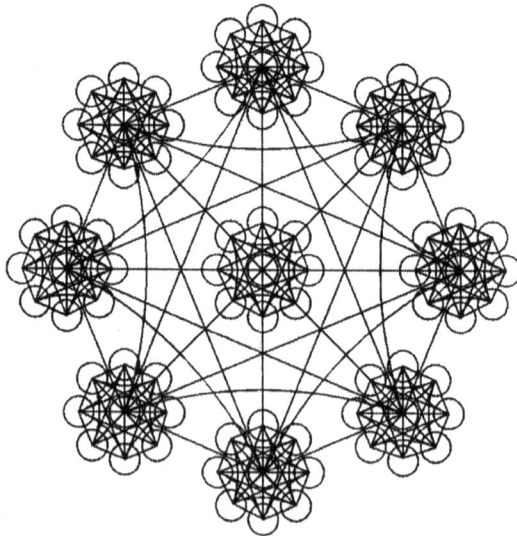

Fig. 10. City. The illustration shows a larger city structure composed of nine city units. In this configuration, each urban community is no more than three transit hops from any other urban community.

It is easy to see how we can extend our nestled design to accommodate large populations on small geographical areas. Say the building block of our city, the urban community, has a population of 50 thousand people. Present day Hong Kong has a population density of about 40 thousand people per square kilometer, or 100 thousand people per square mile, with a lot of space taken up by streets, parking structures, and in other ways not efficiently used. The city unit depicted in *Figure 9* would then have a population of 450 thousand living in an area of 9 square kilometers, or about 3.5 square miles. The city depicted in *Figure 10*, which adds a structural level, would have a population of 4 million people living in an area of 81 square kilometers, or about 31 square miles. In this configuration, we could reach any place in the city by walks and transit hops in about twenty-one minutes.

If we added one more structural level and included four of the city structures depicted in *figure 10*, we would have an urban area able to accommodate the 16 million residents of the Los Angeles basin, now spread out over some 3,200 square kilometers, or 2,000 square miles, depending on where one draws the urban boundary, on 324 square kilometers, or 125 square miles. We could access any part of this urban structure with one additional transit hop, adding three minutes to our travel time. Completing the level by increasing the number of city structures to nine allows us to accommodate 36.5 million residents, more than the population of the Tokyo area or the Mumbai area, with no additional transit hops or travel time. We could get anywhere in an urban area more populated than any that now exists in about twenty-four minutes.

By getting rid of roads and parking lots, we shrink the spatial geometry of the city and create an urban environment structured around population centers as opposed to disperse geographical locations. Like a biological organism, systems of human interaction are nestled within systems of human interaction to form a coherent whole—the city. In addition, our design is geographically independent. The city has a basic spatial unit, the one square kilometer circle of the urban community, but the arrangement of these units is not geographically based. Because we link urban communities with a transit system, and we can run our transit cars at different speeds, we can vary the distances between urban communities without increasing transit times. As *Figure 11* shows, we can superimpose our city over almost any landscape without disrupting our city's nestled structure.

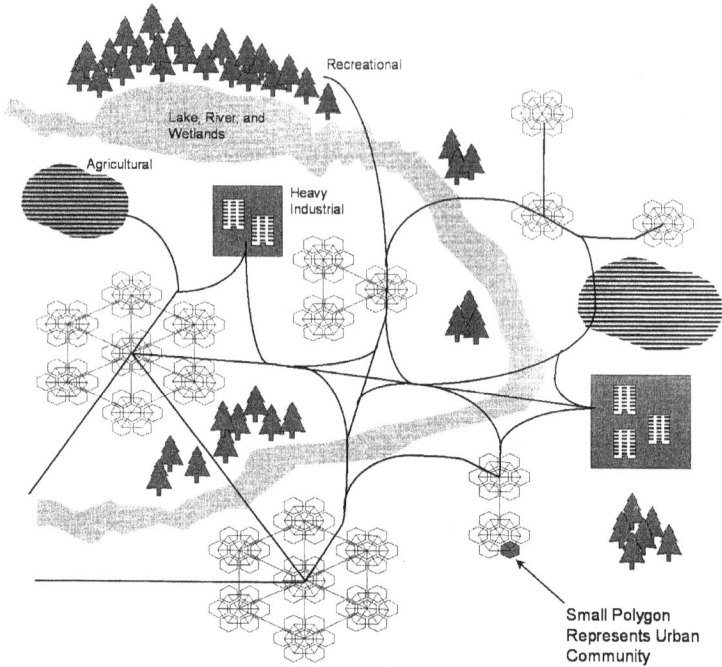

Fig. 11. Urban Landscape. The illustration shows a possible urban layout. Note, the integration of industrial, recreational, and agricultural land. Also note the coherent transportation system and the nestling of urban communities within larger urban regions in a way the will reinforce the less collective societies of the future. In the configuration shown, the city would be home to more than seven million people.

This outlines the framework of tomorrow's city. We have a design that provides for our freedom and mobility as individuals and that accommodates our social structure and its evolution. Now it is time to flesh out our plan, to add details.

We begin with the building block of our city, the urban community. We want our urban community to be a vibrant place but not one that is so crowded we feel cramped. How will we accommodate fifty thousand people, or whatever number we ultimately come up with, on one square kilometer? We start with the bottom half of our urban community—the *foundation structure.*

Every building has a foundation; and, as in rural construction, the foundation of our urban community must be able to withstand any forces we would expect it to be subjected to. In the future, military weapons and industrial accidents will not be the issues they are today, but our cities, and the foundations on which they rest, must be able to weather fires, tornadoes, hurricanes, and earthquakes without damage or interruption of function. Like dams and other grand works of architecture, we must also engineer our urban structures for a long working life, hundreds even thousands of years.

In our urban community, the limiting design factor is earth movement. As we know, the rough shape of the urban community is a circle a little more than one kilometer in diameter, an area eighteen times the base of the Great Pyramid. We, however, cannot build a foundation structure this large and expect it to withstand an earthquake. Earthquake force travels in waves. As an earthquake wave moves through the ground, it would lift one part of the foundation and lower another, like when on a pond a wave passes through a lily pad. There is no practical way to reinforce a rigid structure of this size against the stresses a severe earthquake would impose, so we must make our foundation structure flexible. As *Figure 12* shows, the solution is to divide our one square kilometer circle into chunks that are small and strong enough, and with enough space between them, so that when jiggled around by an earthquake they will not be damaged or bump into each other.

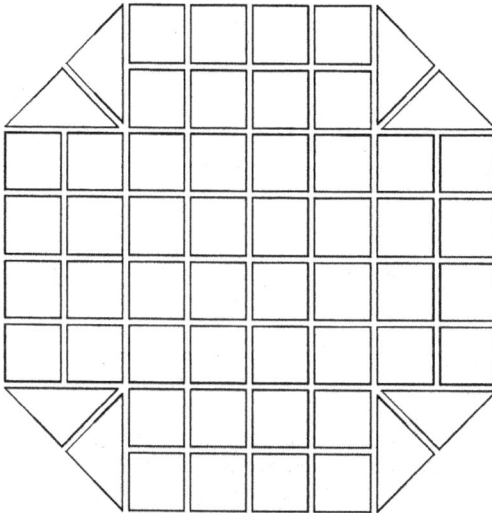

Fig. 12. Foundation Blocks. There is no practical way to reinforce a rigid structure the size of our urban community against the forces imposed by a severe earthquake, so we must make our community's foundation structure flexible. The way to do this is to divide our one square kilometer circle into chunks that are small and strong enough, and that have enough space between them, so that when jiggled around by an earthquake they will not be damaged or bump into each other.

As any engineer will tell you, the best place to build this or any other type of structure is on bedrock. Say that at the site we have selected for our urban community bedrock lies eighteen meters, or about sixty feet, below the surface and that we have excavated to that depth. Now, as *Figure 13* shows, imagine that the independent blocks of our foundation structure extended up from the bedrock to the surface. Also imagine that our structure was near the coast and subject to hurricanes and storm surges and that we extended our independent foundation blocks up an additional eighteen meters. If our building material is concrete reinforced with steel, carbon fiber, or some other material with good tensile strength, we now have a foundation structure able to withstand fire, tornados, earthquakes, and the most severe tropical hurricane.

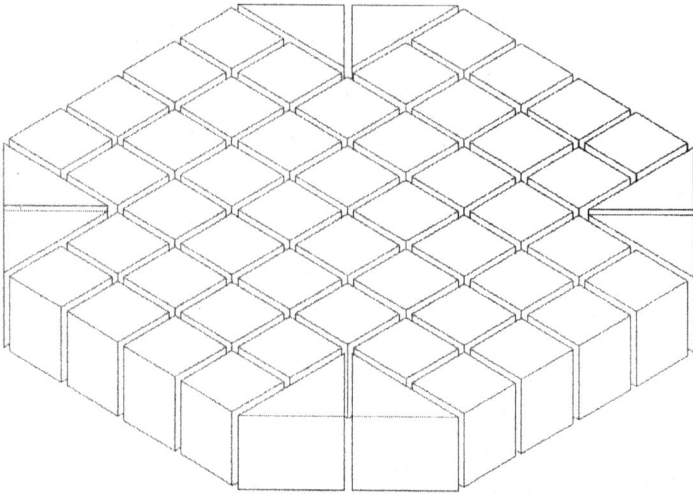

Fig. 13. Foundation Structure. The illustration shows a foundation structure
that extends up from bedrock high enough above the ground to protect
from floods and other natural hazards. In the foundation structure, we
would install our urban community's mechanical systems. On the foundation structure, we would build the community itself.

We also have a twelve-story partially buried structure with a lot of room inside. The lowest levels of our structure would be a good place to put sewage tanks, one or more in each foundation block. We will probably treat sewage in plants outside the urban community, but the urban community would need some ability to buffer sewage flows. Above the sewage level, we would put water tanks. Analogous to the way we deal with sewage, we will probably purify water in an outside plant, but some storage in the urban community would be necessary. Further up, we would have power, communication, and heating and cooling facilities. Above that and above the height of any storm surge, we would run the pipes and cables that link the foundation blocks, designing lines to withstand movement between blocks.

We would also have to incorporate a facility to route garbage out of the community and a facility to route goods into and out of the community. We now have a foundation structure with the mechanical systems to provide water, sewage, power, communications, air conditioning, and waste and goods transport. Systems would be accessible for maintenance and replacement and there would be room for redundancy. When a water tank needs cleaning, we will draw water from other tanks in other foundation structures or in other sections of our structure. Our foundation structure also gives us an easy way to route utilities—vertically to higher levels.

We can think of the foundation structure as a standard component. We may vary the height and size of the foundation structure to accommodate the depth of bedrock or other site consideration, but its dimensions are fixed by human social and physical needs. Within the foundation structure, we can install and upgrade any number of mechanical systems, but there are only so many ways to build the structure itself. Construction involves engineering on a massive scale but in a simple and repetitive way.

The foundation structure may be a standard component, an exercise in engineering creativity. But its uniformity creates the situation where architectural expression can unfold. The foundation structure is the base on which we can build almost any arrangement of structures we can imagine. It is the platform on which we can over time perfect the urban community.

Directly on the foundation structure, we will place levels that house our transit terminals. These levels will also house shops, theaters, classrooms, restaurants, and almost any business we can imagine and that we can conduct in a densely populated urban environment: shoe makers, clothing makers, furniture makers, healthcare providers. Higher up, we will have more levels filled with more shops, theaters, classrooms, restaurants, and businesses. We will mix schools with commercial activities. We will mix businesses with living spaces and recreational facilities. Our imagination knows no bounds. Above the sweeping expanses of our urban community's lower levels, grand towers will soar skyward.

We will anchor these and other buildings to the independent blocks of our foundation structure. This allows the urban community to as a whole flex and thus to absorb earth movement. As we walk the city, however, we will probably not be aware of its compartmentalization. We will bridge the short distances between each of the foundation structure's blocks by pedestrian avenues and by levels of pedestrian avenues. Horizontally, transportation will be on foot. Vertically, transportation will be by stair and elevator. Banks of elevators will service tall buildings. Stairs will service the terraces and walkways of lower levels. Whereas the design of the foundation structure is largely fixed, the design of the urban community will be flexible. On the foundation structure, we can build, tear down, and remodel for centuries to come without affecting our community's structural and mechanical integrity.

The same standards of beauty and functionality apply to our city's personal spaces, to our homes. Unlike today's cramped city apartments, under the economics of fulfillment systems of the future we will not be bound by cost and budget concerns and can shape our living areas in any way we imagine. Constructed in towers with windows and balconies that overlook the city, we will design spaces that are architecturally classic, architecturally modern, and everything in between. Because fire and maintenance are critical concerns, many homes will have marble walls and tile and granite floors. All will have the best workmanship and the best handcrafted cabinetry and furnishings our economics of fulfillment system can deliver.

In every way, our urban community will be an expression of who we are and of where we are headed. A communication system will allow us to interact with anyone anywhere at any time. Transit will operate like horizontally running elevators. To be whisked to our destination community, we need do nothing more than step aboard. City transit will connect urban communities and link the urban community to larger city units and to airports, sports arenas, and industrial facilities. Regional transit will connect cities. Airliners will traverse oceans and continents. A part of the urban canvas we must not ignore are the spaces around the urban communities and between the city units. We will nestle our urban communities in parks and garden plots. We will surround larger urban units with farms and lakes, bike paths and hiking trails. We may also locate rural communities in these areas, increasing the diversity of housing and transportation options in the urban vicinity. The city of tomorrow promises to provide a dynamic urban environment, but one that is no more than a transit hop to the countryside.

The building block of the future city is the urban community. It will function as the core component of the urban framework and as such serve as the basis for a regional, national, and global social structure. The urban community will also support a diverse economy and sustain more intimate levels of social configuration. It will be our place, our home, the geographical area where we will live, work, and feel we belong. We will have our circle of friends, family, and coworkers, in whatever form humankind's evolving social needs will shape these relationships. Just as our rural community promises the best of small town life in the countryside, our urban community promises the best of urban life in the city.

In this section of the book, we established design criteria for the earth's future urban and ecological infrastructure and then applied our criteria to draft a blueprint for reconstruction. The guiding principle of our design was to establish the urban and ecological landscape that allows the individual to exercise his or her creative power and that through its planning and building serves as an avenue for the individual's creative expression. Our countryside has wilderness areas, intermediate zones, agricultural land, and rural communities tied together by a coherent highway system. By way of a transit link, the rural community serves as the access point to the city. Like a living organism, tomorrow's urban landscape will consist of centers of human activity nestled within centers of human activity, integrated into a greater whole by a coherent communication and transportation system.

PART THREE

Implementation

8

Working Drawing

WE HAVE DRAFTED A general blueprint for reconstruction. Now it is time to get our hands dirty. In this section of the book, we look at the issues we will face in implementing our plan. What are the preliminary steps of the construction process? What hurdles will socialism and capitalism pose to reconstruction, and how will we overcome these hurdles? How will we meet our energy needs, and how will we design for the earth's evolving climate?

We begin by expanding our design concept into what engineers call a *working drawing*, or the aspect of the overall plan that addresses the details of construction. What design factors must we address in our working drawing, and what provisions must we make to gather and organize the material and human resources we need to build off of our working drawing? As important, how will we overcome issues of *stagnation* and *fundamentalism*, the battle that we as individuals and as a human community wage against obsolete ways and ideas—the future versus the past?

The first step in drafting our working drawing is to identify a distinct natural zone. The boundaries of this zone would be based on ecological and geological divisions: on coastlines, river valleys, mountain ranges, forest extents, and other natural considerations. Bounded in this way, the zone becomes a cohesive geological and ecological unit, and it will have an optimal design for reconstruction. There will be a best place to bound a wilderness area, a best place to put roads, farms, rural communities, and urban centers.

Though from an overall standpoint we will partition the earth based on natural geological and ecological divisions, we must also take into consideration human divisions. Traditionally, humanity has segregated itself by race, wealth, language, and ethnicity. Our towns have their rich neighborhoods and their poor neighborhoods. Our cities have their Irish, Italian, Haitian, Chinese, Latin American, African American, and other communities. As humanity becomes more interconnected—and as we rise above the financial divisions created by scarcity-based economics— segregation in this way will become less important. We, however, will continue to value our heritage.

English, for example, has become the world's dominant language. It is the common language of the Internet. It is the common language of trade and commerce. It is the common language of science and engineering. As time passes, humanity will adopt a global language, reasonably English or what English evolves to become. We, however, will continue to use local and ethnic languages and dialects. The French will continue to speak French. The Germans will continue to speak German. The Mexicans will continue to speak Spanish, and the Chinese will continue to speak Chinese. People living in the state of Texas will have their dialect of English. People living in Ireland and Scotland will have their dialect. The members of every culture will speak the common language of the globe, their local or cultural language, and whatever other languages they find useful and choose to learn.

Similarly, we will embrace the architectural traditions of our culture. We may build country homes to blend in with and enhance their setting, a style elevated to a high form by the architect Frank Lloyd Wright. In the forests of the Northwestern United States, for example, houses may have the gables and steep roofs of the forest lodge. In the deserts of the Southwestern United states, houses may have the arched doorways and open courtyards of the traditional Spanish home. One can only imagine the cultural expressions we incorporate into the architecture of our urban communities and their soaring towers. Our blueprint for reconstruction carries urban design to a new level of coherence, one that allows for the expression of cultural traditions and the world's great architectural styles.

We will also continue to divide the earth based on states and nations. These divisions, however, will not be grounded in politics and the need to control the population but on the evolution of human social structure from more to less collective forms. In the future, the immediate and extended family, and our circle of friends and associates, will become more important. Such structures will serve as the foundation on which we build the community, which will serve as the foundation on which we build the

city and so on, up to the level of the state and nation. As we described in
Threshold to Meaning: Book 2, Economics of Fulfillment,[1] such governmental structures will be less intrusive into our lives than they are today.
Rather than control the individual, government will serve to facilitate the
creative evolution of the individual. Government will reflect and allow
for, not impose, social order.

Our embedded levels of social structure will also give rise to a global
social structure. This social structure, however, will have little in common
with the ideas of global government embraced today and lumped under
the heading of "new world order." There are many variations on what,
today, we think global government should be. All, however, are based on
the idea of a global government as a mechanism of central control with
jurisdiction over the nation. Variations range from the standardization of
environmental and health and safety regulations across national boundaries to socialistic ideals of totalitarian control over subservient bodies.
Many in the environmental community see global government as a way
to regulate every aspect of the individual's life for the good of the planet,
as they define what is in the planet's interest. Many have also latched onto
environmental and globalist ideology to further objectives of personal
and institutional power. A global social structure will emerge in the future; but, as with the nation and state, it will exist to facilitate rather than
regulate the creative activity of the individual. The global government of
tomorrow may be little more than a mechanism to help us coordinate
reconstruction on a planetary scale, one that arises when needed and that
vanishes when no longer useful.

In the future, we will partition the earth based on local, state, and
national boundaries, and we will create some form of global coordi-
nating mechanism. Human divisions, however, will be incorporated
within geological and ecological divisions. The Columbia River in North
America, for example, originates in Canada's British Columbia prov-
ince and, in the United States, travels through or drains territory in the
states of Washington, Oregon, Idaho, Montana, Wyoming, Utah, and
Nevada. These artificial boundaries may continue to exit—we may still
have a Canada and a United States, a Washington, an Oregon, an Idaho,
a Montana, a Wyoming, a Utah, and a Nevada—but we will plan for the
Columbia drainage from its source in the Canadian Rockies to its mouth

---

1. See *Book 2: Economics of Fulfillment*, chapter 10.

in the Pacific Ocean. Whatever human divisions we impose on the landscape, they will contribute to the larger objective of earthly perfection.

With a geological and ecological zone identified, our working drawing must address the specific issues of construction. In the countryside, we must identify wilderness areas and determine where to place boundaries between wilderness categories. What land do we manage as traditional wilderness, and what land do we manage for safety and access? We must delineate intermediate zones and map timber, mineral, recreational, and other resources. Where can we dig a mine? Where can we create, sustain, or remove a forest? Where can we build a hydroelectric dam, or where can we remove one? In agricultural areas, what soil types and other conditions do we have to work with? What are the stream flows, and where can we build irrigation systems? Where will we locate rural communities, and what will be their layout and the design of their structures? How will we run roads, and where will we place automobile support facilities? Will we build new cities over old cities or on undeveloped terrain?

In the city, the basic design component, the urban community, will rise on a circular plot of land with an area of one square kilometer. It consists of a foundation structure on which we will erect the community itself, home to fifty thousand people. Is this a reasonable population density? To answer this question, builders must calculate our public and private space needs. How much living space will we each want: one thousand square feet, two thousand square feet, three thousand square feet? How much business and public space will we each want? Will the one square kilometer circle of our community enable us to meet our space requirements? If we were to increase our circle's diameter to one mile, we would extend our walking time by only four minutes but more than double our circle's area. This would allow us to accommodate twice as many people or to cut population density in half. In some cultures and ecological zones, we might prefer a higher population density. In others, we might prefer a more spacious urban plan.

As important, we would have to address design issues in the urban community itself. Commercial and other space would have to be accessible without congestion. Walkways, stairways, and elevators would have to accommodate use. Buildings would have to withstand the most severe natural hazard. How tall will we build skyscrapers in the future? From a structural standpoint, we can make them much higher than we do today, but there is no practical reason to do so. As height goes up, we must add elevators. If we make our building too high, we lose more interior space to the extra elevators we must add than we gain on the extra floors.

We must also address design considerations in planning our transit system. Most of our daily activities will take place in the urban community, but a certain percentage of the population will travel between communities, and our transit system must be able to accommodate the load. Reasonably, each transit line would lead to and from a single terminal. This would eliminate the need for stops on the way and would work well with our city's nestled design. The speed of the transit link would vary with the distance traveled. Transit cars on short hops will travel relatively slow. Transit cars on long hopes will travel faster. How fast will transit cars between cities operate: 200 kilometers per hour, 300 kilometers per hour, 400 kilometers per hour? We cannot run transit lines overland where, like today's rail systems, they would cross roads and meet up with cars, people, and animals but would have to elevate them or run them below ground. And, to erect transit lines on a massive scale, we would have to engineer them in a standard way—mass produce towers, tracks, bridges, and transit cars.

When we know what to build and how and where to build it, our task becomes to organize the resources to get the job done. As we discussed in chapter 5, *Design Criteria*, contemporary economic systems are not structured to allow us to coordinate our efforts on the grand scale of global redesign. In today's scarcity-based, socialist-capitalist world, our dream of global reconstruction would not be practical. It would cost too much. Construction and resource acquisition will take place through the economics of fulfillment framework. Economics of fulfillment is the tool of reconstruction. It allows us to put money and budgets behind us and align our energies to a higher end.

The first economics of fulfillment systems will exist alongside scarcity-based systems. We will import resources from the outside, socialist-capitalist economy and bring outside businesses under the economics of fulfillment umbrella—steel mills, rock quarries, cement plants, heavy equipment manufactures. As important, we will build our own facilities. We will open businesses and build the plants and equipment needed to move vast amounts of earth, to transport, pour, and cool vast amounts concrete, and to cut and polish vast amounts of marble and granite.

As we engineer our landscape, we will empty huge tracts of formerly built up land. How will we dispose of asphalt and concrete? How will we reclaim steel reinforcement from demolished bridges and buildings? Scarcely will we have a need for timber in the years ahead. As we tear down subdivisions and reclaim the farmland on which many were built, we will make available billions of board feet of lumber now tied up in homes and strip malls. Under the economics of fulfillment model, we will establish the businesses that enable us to build on and creatively discard the old to create the new. Through these businesses, owners will fulfill their creative needs.

As we also discussed in our chapter on design criteria, our economics of fulfillment philosophy is the tool that provides the human resources for reconstruction. How many people will it take to build a city of thirty million? Economics of fulfillment aligns our need for creative expression with our task of earthly perfection. As such, the labor force to rebuild the earth's urban and rural infrastructure is not something we must hire. It is something we are and that by our every action we participate in. As diverse as our dreams and ambitions may be, our acceptance of meaning—our grasp of a common future—aligns our desires toward a higher end. We are the human element of reconstruction. Labor is the human community. We achieve self-perfection through earthly perfection.

In drafting our working drawing for reconstruction, we will also face conflict between dreams of the future and ideals of the past. To place a highway or build a ski resort, we will have to overcome economic, historical, environmental, and other political concerns that have lingered to the time of reconstruction.

No optimal plan will call for clear-cutting a wilderness area, though, as in chapter 6, *The Countryside*, we discussed, clear-cutting can be a valuable forest management tool. We will not remove the forests around Oregon's Crater Lake or in Montana's Glacier National Park, but we will dam rivers, build roads, harvest timber, and extract minerals. And there will be those who will, driven by an environmental doctrine that considers human activity to be incompatible with their goal to return the earth to its "natural" state, oppose our actions. This conflict, however, may not be as significant as we may think. Foresters have long sought better access to the woods, and moderate environmentalists have long sought the sustainable harvest of resources. As we replace the ideologically driven environmentalism of today with the evolution of consciousness driven environmentalism of tomorrow, we will absorb the extremist movement of old and end the conflict created by environmental fundamentalism.

Just as no optimal plan will call for clear-cutting a wilderness, no optimal plan will call for tearing down Rome or Saint Petersburg's architectural treasures. But we will build New London over old London, New San
Francisco over old San Francisco, and our working drawings must make
provisions for historical preservation. We will not latch onto historical
preservation with the religious zeal embraced by many who populate today's historical commissions—of those who in the name of tradition lock
in the engineering inadequacies of the past. We will embrace historical
preservation in a practical way, to a degree that preserves the past without
sacrificing the future.

Similarly, our working drawing must take into account those who
profited from the economic practices of old. In today's socialist-capitalist
world, capital equates with power. In the future, there will be those who
cling to their economic position and to what real or imagined qualities they feel it brings to their lives. But what value is money when you
have what money can buy? What value is power when you have the opportunity to create and to express your ideal of perfection? Tomorrow's
leadership will be rooted in vision and substance of character and not in
greed, money, and politics.[2] As economics of fulfillment takes hold, the
dynamics of its growth and the personal and social evolution it nurtures
will overcome the economic fundamentalism of the past.

Economics of fulfillment will also enable us to move beyond the legal
framework of the scarcity-based world. Take the concept of work and the
entanglement of wage-and-hour laws it embodies. In a world where life
in not a struggle to survive and we are not a component of the production
process—the labor part of land, labor, and capital—the notion of work
has no meaning. We may work eight-hour days, ten-hour days, sixteen-
hours days. We may contribute to earthly perfection by taking time off
to spend with our children or to gather our thoughts. Creativity is fluent,
dynamic. The job of the steelworker is no more important than that of
the artist. Or are we all artists? Our task of earthly perfection will not be
our work. It will be our life. Work is only work when we would rather be
doing something else and there is something more important to do. In today's world, a legal framework mediates the actions of one faction against
the other; our economy cannot function without it. In tomorrow's world,

2. Ibid., chapter 7.

factions and individuals will not be at odds, and the legal framework of the past will wither into history.

For the geological and ecological zone we have identified, our working drawing represents the optimal design. It transforms our vision of tomorrow's world into the measured plans we as builders use on the job. It is the tool that allows us to implement our ideal of perfection. As in any creative endeavor, our plan must be flexible. It must evolve as our ideal of perfection and our understanding of how to create perfection evolves. Our working drawing represents the ever-refining details of construction.

9

Energy

TODAY, A SOURCE OF clean and abundant energy is one of humankind's most pressing concerns. In the pages ahead, we explore this issue. In that this is the third book in the *Threshold to Meaning* series, and that our evolution of consciousness view embraces a fundamentally different conceptualization of atomic and subatomic structure, we will provide more than the typical discussion on energy use and creation. What is the essential nature of energy? Must energy be "conserved," as physicists tell us, or can energy be created? If so what is the physics behind energy creation? What are our traditional sources of energy, and how do we harness these sources? What nontraditional sources of energy might we develop in the future? Will our solution to humanity's energy problem have anything at all to do with a new source of energy? Perhaps we will find the answer to our energy problem somewhere else, in front of us all along.

Typically, we think of energy in terms of its effect on matter. There is potential energy, or the latent ability to do what physicists call work. There is kinetic energy, or the energy of movement. There is thermal energy, chemical energy, mechanical energy, electrical energy, and nuclear energy. Though it is useful to think of energy in traditional ways, in our evolution of consciousness view, energy is not as we have been taught.

As we described in *Threshold to Meaning: Book 1, Evolution of consciousness*, energy is generated in the creative cycle.[1] Energy is the consciousness of need, the awareness of desire. It is motivation. It is the urge to create satisfaction, the drive to create fulfillment. From the hu-

---

1. The explanation of the creative process as it pertains to energy formation included in this chapter deviates from the standard view of energy put forth by physics. For a better understanding of the creative nature of energy, see the first two sections of *Book 1, Evolution of Consciousness*, where the creative process is developed and justified.

man vantage, however, energy reveals itself in different ways on different evolutionary levels. On the subatomic scale, we see energy as the *weak* and *strong* forces that hold together the atomic nucleus. On the atomic scale, we see energy as the *electromagnetic* force that supports atomic structure and that allows atoms to form molecules. On the cosmic scale, energy reveals itself as *gravity* and *inertia*. In the realm of life, energy takes the form of thought, learning, and organic reproduction. In the human realm, it reveals itself as our drive to understand and our yearning for the freedom to perfect ourselves and our surroundings. There is only one energy—one force, one drive, one motivation—and it is a dimension of consciousness. This energy, however, manifests in different ways in the physical forms that represent different stages of evolution.

Neither, in our view, is energy necessarily conserved. Armed with Newton's laws of force and inertia and with more recent laws of matter and energy conservation, physicists believe that energy can change form but not be created or destroyed.[2] A battery can convert chemical energy into electrical energy; but, given ideal efficiency, every joule of chemical energy lost must be matched by an equal number of joules of electrical energy gained. The big bang model of cosmic formation states that all the energy now in the universe, and that ever will be in the universe, existed at the universe's beginning. As we described in *Book 1, Evolution of Consciousness*, the energy generated in the creative cycle increases depending on the extent of the *autocatalysis of consciousness* that takes place in the cycle.[3] In this respect, our concept of energy violates the conservation laws of physics. Or does it?

To see how our idea of energy ties in with physics and its conservation laws, we need to look more closely at the creative process. The creative process builds on and creatively discards the old to create the new. As the leading arrow of the universe's evolution jumped from subatomic structure to atomic, molecular, cosmic, and organic structure, each preceding stage of advance fell into the realm of evolution's trailing arrow. Atomic, molecular, and cosmic activity realigned in support of the universe's overall forward movement. This situation created the predictable world physicists love. Creative activity on evolution's trailing arrow progressed in ways that were consistent, in ways that physicists could map

2. In an expanding universe, and with regard to recent theories of dark matter and dark energy, physicists concede that the law of energy conservation does not apply.

3. See *Book 1, Evolution of Consciousness*, chapter 3.

with such regularity that they took them to be laws. Energy is generated in the creative cycle; but, in a typical chemical or nuclear reaction, it does so in a way where from the external vantage of physical measurement energy appears to be conserved.

Take the forward and reverse reactions that describe the synthesis of oxygen and hydrogen to create water and the breakdown of water to create oxygen and hydrogen.

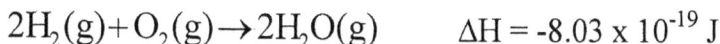

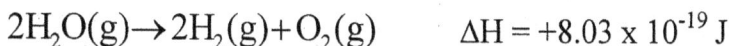

$$2H_2(g) + O_2(g) \rightarrow 2H_2O(g) \qquad \Delta H = -8.03 \times 10^{-19} \text{ J}$$

$$2H_2O(g) \rightarrow 2H_2(g) + O_2(g) \qquad \Delta H = +8.03 \times 10^{-19} \text{ J}$$

The first equation states that two hydrogen molecules, H, combine with one oxygen molecule, O, to form two water molecules, H_2O, releasing 8.03×10^{-19} Joules, J, of energy, ΔH. The second equation states that two water molecules, H_2O, break apart to form two hydrogen molecules, H, and one oxygen molecule, O, consuming 8.03×10^{-19} Joules of energy. The customarily added letter in parentheses indicates that molecules are in the gaseous state.

If, in the forward reaction, we let the arrow denote the middle of the equation, we have three simple molecules on the left and two complex molecules on the right. The reaction thus proceeded from a state of lower complexity to a state of higher complexity. External complexity reflects internal complexity, which reflects consciousness. By creative process, the three aspects of lower consciousness and less complex structure of perception on the left side of the formula crossed a threshold to break apart. Their component atoms then recombined into two aspects of greater consciousness and more complex structure of perception. In the process, the entities of consciousness that comprised the initial relationships moved from a less fulfilled to a more fulfilled state. Correspondingly, their manifested structures moved from a high to a low mass-energy state, with the difference measured as a release of energy. In the reverse reaction, the opposite sequence of events takes place. The products, oxygen and hydrogen, are in a higher mass-energy state than the reactant, water. To form the product we would therefore have to input energy. The reverse reaction is what chemists call *endothermic* as opposed to *exothermic*.

Although the creative cycles that underlie the forward and reverse reactions produce energy, from the standpoint of the reactions taken together energy appears to be conserved. As much energy is created in the forward reaction as is used to initiate the reverse reaction, or to generate the state of evolution where the reaction can take place. Actually a little more is used because the second law of thermodynamics tells us that some energy would be lost to entropy, or to inefficiencies in the reaction. The net direction on evolution's trailing arrow is toward decline.

There is only one fundamental energy, the need to create fulfillment, the drive for perfection. From the practical standpoint of developing a source of physical energy, however, we can think of energy as associated with two levels of physical behavior. Between thresholds, energy is generated in the creative cycle. We will call this level of physical behavior *intra-threshold* physics. Across thresholds, or in common chemical and nuclear reactions, energy is conserved. As much energy is created or expended on one side of the threshold as is created or expended on the other side. We will call this level of physical behavior conventional, or *inter-threshold,* physics. Our idea of the creative process and our notion of energy creation have not broken the laws of physics. We are describing a level of physical behavior where those laws do not apply, a dimension of the physical world contemporary physics has only begun to explore.

Now, let us apply what we know about energy to the task of harnessing it. For the sake of familiarity, we will begin with traditional energy sources, those that obey the laws of conventional, or inter-threshold, physics and that today power our cars and light our homes. All the energy we at present harness in some way comes from two sources: the sun and the earth.

Solar energy drives winds and ocean currents. It powers hurricanes and thunderstorms, and it lifts water from the oceans and drops it onto the land where it runs back into the oceans. Through the process of photosynthesis, plants use solar energy to synthesize carbon and hydrogen into sugars, amino acids, and carbohydrates. Animals, and those plants such as fungi that do not conduct photosynthesis, consume photosynthetic plants or those organisms that do and indirectly harness the sun's energy. On every level of the biosphere, life is driven by the sun.[4]

---

4. We are disregarding life that exists in thermal hot springs on the ocean floors and other areas that are out of reach from the sun. For the most part, life on earth is sustained by solar energy.

And, wherever we can do so practically, we tap into the flow of solar energy that strikes the earth and that works through the earth's various systems. Photoelectric cells convert radiant energy into electrical energy. Other types of solar panels use radiant energy to heat water and to warm homes. Windmills and dams with turbines convert the movement of air and water into mechanical energy, which we use to run generators that produce electrical energy. In agriculture, we use photosynthesis to lock solar energy into grains, fruits, and vegetables and, indirectly, into beef, pork, and chicken. We refer to sugars and other organic materials as *hydrocarbons* because of their hydrogen-carbon based molecular structure, and we turn hydrocarbons into fuel. When fermented, grain and other organic materials produce methane, which is the primary constituent of natural gas, and alcohol, which, because it burns without leaving as many damaging deposits as gasoline and ignites more controllably, is the fuel of choice for high performance, high compression racecar engines.

Our most important source of solar energy, however, does not come from sunlight that strikes the earth today. It comes from sunlight that struck the earth long ago. As we have learned, the biosphere was more complex in the past than it is today. Life was more diverse and plentiful, and it locked up vast amounts of solar energy in its organic structure. We, today, tap into the hydrocarbon remains of ancient life and harness the solar energy they contain. In the *Carboniferous Period* of the Paleozoic Era, for example, the organic material from vast fern forests accumulated over millions of years to form many of the world's large oil and coal deposits. Today, these "fossil fuels" are our primary source of energy. When burned, coal, propane, gasoline, jet fuel, and heating oil release solar energy trapped long ago in organic matter and made available to us as hydrocarbon remains.

The sun is humankind's primary source of energy, but we also derive energy from the earth. As we know, the earth is hot inside, its temperature in part the result of the decay of uranium and other radioactive elements. Although the heat inside the earth is a seemingly inexhaustible source of energy, it is hard to get to. We can only harness *geothermal energy* in a substantial way in areas where temperatures are warm close to the earth's surface, in volcanic regions or regions near major fault lines. In the southern part of the state of Oregon, for example, the Oregon Institute of Technology is heated by geothermal energy, as are many buildings in the nearby city of Klamath Falls. In Iceland, geothermal energy heats the city of Reykjavík, provides steam to industrial facilities, and is used to heat greenhouses for vegetable production. Iceland also uses geothermal energy to produce hydrogen for fuel.

More commonly, we draw energy from the earth by harnessing the nuclear reactions thought to heat it, for example the decay of the Uranium-235 isotope into rubidium and cesium:

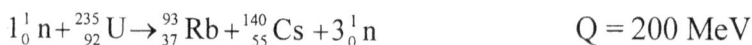

$$1_0^1 \, n + {}_{92}^{235} U \rightarrow {}_{37}^{93} Rb + {}_{55}^{140} Cs + 3_0^1 n \qquad\qquad Q = 200 \text{ MeV}$$

The above expression states that, when struck by a neutron, one uranium-235 atom, U, decayed to form one Rubidium-93 atom, Rb, one cesium-140 atom, Cs, and three neutrons, n, releasing 200 millielectron volts, MeV, of energy, here indicated by "Q." The energy released in a nuclear reaction is normally expressed in MeV; 1.0 MeV equals 1.60 x 10^{-13} Joules. The other figures associated with each element indicate its atomic number.

Scientists once believed that this type of nuclear reaction, called a *fission* reaction, would provide the world with an inexhaustible source of power. The pioneers of fission, however, never envisioned their technology's political consequences. In this regard, a brief look at fission and its ramifications will be useful, because the technology will play a role in our blueprint for reconstruction.

As you may recall from your high school or college physics courses, an atom has a nucleus made of protons and neutrons. Nuclear fission takes place when the nucleus of a complex atom breaks apart to form the nuclei of less complex atoms. In the process of this *decay*, energy is released—a lot of it, up to ten million times as much as in a chemical reaction.

In the area of power generation, the most common reaction is the decay of the uranium-235 isotope. As the equation shows, the reaction is initiated when a neutron strikes the nucleus of the uranium-235 atom and results in the release of energy and the production of Cesium-140, Rubidium-93, and three additional neutrons. These neutrons then go on to strike other uranium-235 atoms, which causes a chain-reaction and a continuous release of energy.

To regulate this chain reaction, a reactor is designed to control the flow of neutrons. This is accomplished by separating uranium fuel rods with a *moderating medium* and by removing or inserting neutron-absorbing materials between fuel rods. Carbon, sodium, hydrogen, deuterium, and other substances have been used as moderating mediums. In the United States and other developed nations, most reactors use highly purified water.

These *light water reactors* are tried technology and have many advantages. But they have a drawback that in today's world is important. They require a more refined fuel than natural uranium. Natural uranium contains only a small amount of uranium-235, 0.71 percent, with the rest made up of non-fissile uranium-238. To be used in a light water reactor, natural uranium must be *enriched*, or refined to between three and ten percent uranium-235, depending on the reactor's design. The problem with this is that uranium enrichment facilities can also be used to create the uranium-235 ratios necessary to work in a weapon. This is the concern when on the news we hear that gas centrifuges and other uranium enrichment machinery have been detected in Iran, North Korea, or other nation that may want nuclear power for reasons other than to generate electricity.

Light water reactors also have a functional disadvantage. The decay of uranium-235 produces a remarkable amount of energy in comparison with the burning of a fossil fuel but, from a nuclear standpoint in a light water reactor, is inefficient. A typically nuclear plant will only harness about one percent of the energy content of the uranium—and uranium-235 deposits are not inexhaustible. By some estimates, United States reserves could only provide thirty percent of electrical power needs for the next fifty years. Other estimates are higher.

But, although uranium-235 is comparatively rare, it can be used to produce an abundance of other nuclear fuels. This takes place in what is popularly called a *breeder reactor*. When excess neutrons are absorbed by a *fertile* material that material is changed, or transmuted, into a fissile material, typically uranium-238 into plutonium-239. In this reaction, the plutonium then absorbs a neutron and breaks apart, which releases energy and neutrons and produces more plutonium.

To maintain this reaction, a breeder reactor typically uses a molten metal as its moderating fluid, often liquid sodium which melts just under the boiling point of water. A breeder reactor plant can operate as high as seventy-five percent efficiency while producing twenty percent more fuel than it consumes. The first power-generation breeder facility was built in France in 1984. Others have been built in Russia and Great Britain. Today, so much uranium-238 is recoverable from waste and the enrichment process that existing stockpiles could support a full-scale deployment of breeder reactors for centuries.

Nuclear technology has been with us for a long time. It is well understood, and there is no question that it can provide as much energy as we could use well into the future. But there is the issue that makes nuclear power controversial—safety.

In part, this is because nuclear power plants need and create materials that can be used in nuclear weapons. When the United States and the Soviet Union were the dominant nuclear powers, the possibility of engagement was perhaps not as great as we may have been led to believe. These nations understood the power of their weapons and, one could argue, knew better than to use them. The factions bent on acquiring nuclear weaponry today, however, did not share in its development and may have no real understanding of what it can do. Case in point, the Islamic terrorist who in 2007 was caught smuggling weapons-grade uranium in his pocket. As important, these factions often have no geographical base and as such are not susceptible to the threat of nuclear retaliation. Whom would the United States launch against?

The politics of nuclear power are also due to the possibility of an accident. Thick concrete walls shield reactor cores. Plants have backup cooling systems and house reactors in steel and concrete containment buildings. But even with redundant safety features, accidents have happened. In 1979, the Three Mile Island plant in Pennsylvania lost its cooling system. When the backup system failed, the core overheated and dumped radioactive materials into the containment building and a small amount into the atmosphere. In 1986 at Chernobyl in the Soviet Union, one of four early-technology reactors that lacked any form of containment building exploded, contaminating the surrounding area and spreading radioactive material over northern Europe. But is nuclear power as dangerous as opponents of the technology would like us to believe? In the early days of America's nuclear research, seven people are believed to have died from radiation exposure. According to the United Nations, approximately one hundred people died, or are expected to have shortened lives due to cancer, as a result of the Chernobyl accident, a figure seconded by a number of independent investigating bodies. In the United States and Western Europe, no one has ever died or been injured due to radiation in a nuclear power plant and no plant has emitted radioactive materials of any consequence. In contrast, thousands die each year in oil field and coal mining accidents, tens of thousands die from oil and coal related air pollution, and hundreds of thousands die from energy related political unrest.

A nuclear power plant is also seen as a target for a terrorist attack. The reactor in a nuclear power plant, the part that contains the radioactive material, though, is small and easily hardened. Containment buildings are heavily reinforced, and even those built decades ago were designed to withstand bombs and plane crashes. If a cooling tower or other exposed equipment is hit, the plant simply shuts down until repairs are made. So resistant to terrorism are nuclear facilities that, behind the scenes, security experts hope that if a terrorist targets anything it is a nuclear facility. The damage and loss of life would be far less than for other targets.

Also misunderstood is the issue of waste. Spent fuel from light water reactors contains almost all the original uranium-238 and about one-third of the original uranium-235, which can be recycled. The main reason waste has accumulated is that due to political factors nations have not developed breeder and reprocessing programs. By some estimates, a family of four using breeder generated electricity over a period of twenty years would produce no more than a few grams of nuclear waste. The actual danger of nuclear waste has also been overstated. Most of the radioactivity in waste is produced by a few hot isotopes that have short half-lives. Typical nuclear waste will "cool" to the point where it can be easily handled and poses little danger in about fifty years. The present solution to waste is onsite storage at nuclear facilities, though attempts have been made to create centralized storage facilities in stable geological formations, Yucca Mountain in Nevada. Both approaches make waste available for future breeder use and reprocessing, which not only reduces the volume of these materials but puts them to use.

Different nations perceive safety differently, and technology has advanced from the Soviet and Three Mile Island era. Sweden and Austria have limited or terminated their nuclear programs. In the United States, no new nuclear plants have been ordered since 1978; and, at an estimated cost of 100 billion dollars, many completed plants have not been allowed to operate. On the other hand, Japan, France, Germany, and Great Britain have active nuclear programs but face stiff environmental opposition. Current instability in the Middle East, however, has encouraged nations to reconsider the nuclear power option. Canada has plants in the works, and China's continued economic growth hinges on the deployment of nuclear power. Even in the United States, utilities have submitted applications for nuclear facilities. In the current political climate, however, it is estimated that it will take approximately ten years for companies to exhaust the legal and legislative hurdles environmental groups have vowed to raise.

As controversial as it is, given limits to oil reserves and growing political and military tensions over oil, nuclear fission will almost certainly play an increasing role in our future. Nuclear fission also has a role to play in our blueprint for reconstruction. No other source of energy is clean and abundant enough to smelt the vast quantities of steel and produce the vast quantities of portland cement we will need to rebuild our cities and countryside. Despite what the environmentalists may contend, nuclear power is here to stay and in the years ahead will occupy an even more vital place in our lives.

Fission aside, the source of energy scientists at present tout as the ultimate solution to our energy problems is another type of nuclear reaction, this one called *fusion*. Atomic fusion occurs when two atoms with a simple structure combine to create one atom with a complex structure, creating a lower overall mass-energy state and releasing energy. In theory, atoms of any element can fuse to create atoms of a more complex element, but it takes so much energy to initiate a fusion reaction between complex elements that the only practical reactants are the heavy isotopes of hydrogen, in particular deuterium which is extractable from water.

$$ {}_1^2 H + {}_1^2 H \rightarrow {}_2^3 He + {}_0^1 n \qquad\qquad Q = 3.27 \text{ MeV} $$

This equation tells us that two deuterium atoms fuse to form one helium atom and one neutron, releasing 3.27 millielectron volts of energy. The number at the bottom of the element symbol, of course, represents protons and the number at the top neutrons.

In theory, fusion can work without producing the toxic byproducts of fission, but because the temperature needed to initiate a fusion reaction can be in excess of fifty million degrees Celsius, scientists have thus far been unable to sustain a fusion reaction for any practical length of time. It remains to be seen if we can overcome the tremendous technological hurdles necessary to make fusion a practical source of energy.

There is one other source of conventional energy we must look at, the reason is that it is widely misunderstood. This source of energy is hydrogen. It is trendy to think of hydrogen as the fuel of the future, the world's next great well of energy. And, as a fuel, hydrogen has many advantages. It has a high fuel value: 142 kilojoules per gram as opposed to 48 kilojoules per gram for gasoline; and, as our earlier equation shows, it burns to produce water, a byproduct with no adverse environmental effects. What the proponents of hydrogen neglect to tell us is that there is no significant source of chemically unbound hydrogen on the planet. There is a lot of hydrogen, but it is bound in water and hydrocarbons, which means that to release it we must input energy. Moreover, as our example on the nature of energy illustrated, it takes more energy to break apart a water molecule than is released when its constituents, oxygen and hydrogen, are put back together. The same is true of the hydrogen we strip out of a hydrocarbon. We get more energy by burning the hydrocarbon than by breaking apart the hydrocarbon and burning the hydrogen we extract.

For this reason, it is best to think of hydrogen not as a source of energy but as a way to store energy. Solar or wind power, for example, can be used to extract hydrogen from water, and the hydrogen later burned to generate electricity. At the sunniest and breeziest locations, however, solar and wind generators produce electricity only about eight hours a day. Given these limitations, the hydrogen fuel cell cars said to be the wave of the future will do little more than shift the use of hydrocarbons from the automobile to oil and coal fired hydrogen extraction plants and necessitate a greater reliance on nuclear fission.

This look at the common sources of conventional energy brings us to our next topic, unconventional energy, or sources based on intra-threshold physics and that circumvent the conservation laws. Is this where we will find the solution to our energy problem?

To understand intra-threshold physics as it applies to the production of energy, refer to *Figure 14*, which illustrates the cycles of the creative process and the crossing of a threshold.

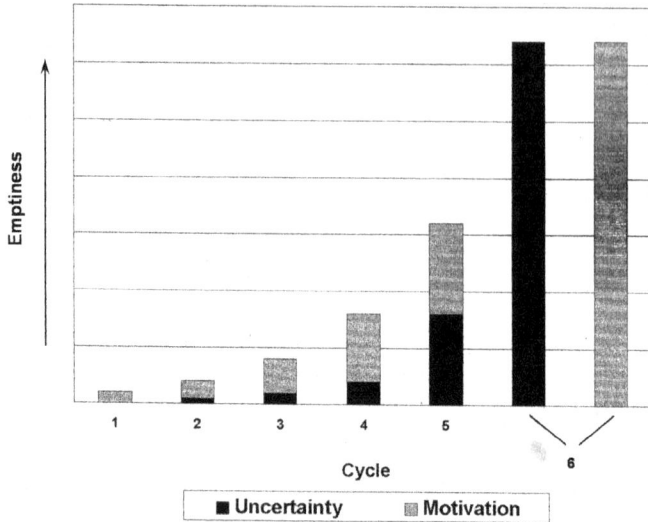

Fig. 14. Energy Creation. When uncertainty reaches the point where it occupies the entire consciousness of emptiness, the exhaustion of the present
evolutionary level becomes certain. Uncertainty collapses and releases the
energy it held at bay, as shown by the second bar associated with the six cycle.

Each bar represents a hypothetical creative cycle. The height of each
bar shows total energy, or total consciousness of emptiness. The lighter
area at the top shows the portion felt as motivation, and the darker area
at the bottom shows the portion felt as uncertainty. With each cycle, motivation increases but, because uncertainty occupies a larger part of the
total, at a decreasing rate. As the last two bars show, when uncertainty
collapses, this energy is released to drive the system across the threshold
to a new evolutionary level, to a new state of existence. What, however,
would happen if we could draw off the energy as it is being released and
prevent the threshold from being crossed? Something useful to say the
least. We would stabilize the creative process in a way that would allow
us to perpetually draw off the energy produced by the creative cycles. The
creative process would function as a generator, as a source of power.

For a more complete explanation of the creative cycle and threshold,
I once again refer you to *Book 1, Evolution of Consciousness*.[5] As for the

5. See *Book 1, Evolution of Consciousness*, chapters 1, 2, 3, and 4.

present discussion, is it possible to tap into the intra-threshold realm of existence? For centuries, scientists have studied the physical behavior that takes place across thresholds. Our model of the creative process suggests that we interact with the intra-threshold realm of physics in a more limited way, at the threshold itself—at the moment when one oxygen and two hydrogen atoms suddenly become water, at the instant when uranium-235 suddenly becomes Rubidium-93 and Cesium-140. This realm of investigation, however, has not been a traditional area of concern in physics or chemistry and as one might imagine is quite difficult to implement. It is at this point however—the flashpoint of transformation—where we may be able to tap into the intra-threshold behavior of matter and energy.

In light of the evolution of consciousness view, the issue is not whether intra-threshold physics and nontraditional sources of energy exist. It is whether we can learn to harness this aspect of nature. Scientists have tinkered with schemes to circumvent the conservation laws since the days of Nicola Tesla, who was one of the pioneers of electrical theory, with no success. Perhaps the theoretical base of intra-threshold physics we established will allow science to approach the problem in a way that leads to a practical outcome.

We have traditional and nontraditional sources of energy. But is either the answer to our energy problem? We have long sought alternatives to oil and coal and long dreamed of the unlimited source of "free" energy that our intra-threshold physics tells us we can, at least in theory, harness. But is any source of energy, traditional or nontraditional, the solution to our energy problem? Perhaps the answer is altogether different. Perhaps we have already found a way to solve our energy woes.

The answer to humankind's energy problem is not the development of a new energy source. It is a decrease in the use of energy—but not in the carbon-offset, bike-to-work way environmentalists proclaim to be our energy future. In our blueprint for reconstruction, we accomplish this objective in a way that works, in a way that we can live with. We will design country homes that function as passive solar energy collectors and radiators. We will design highways for efficiency and serviceability. Our greatest energy savings will take place in the city. We have eliminated the automobile and engineered an urban design where primary transportation is on foot. It will take a great deal of energy to construct the rural and urban infrastructure of tomorrow. Nuclear power will certainly be required. But, once in place, our way of life will demand so little energy that we can provide for our needs through any number of sources: wind, solar, geothermal, the production of methane from sewage, the production of alcohol from farm waste, the production of diesel from food waste. Hydrogen may play a role. Fossil fuels may play a role. Nuclear fission will certainly play a role. We may even perfect nuclear fusion and nontraditional, intra-threshold, sources of energy. But our future will not depend on these advances. The answer to today's energy problem is human evolution. It is our perfection of the earth's urban and ecological landscape.

10

Climate

A S WE HAVE DETERMINED, we must engineer the basic unit of the future city, the foundation structure of the urban community, to have a usable life measured in hundreds even thousands of years. How, over the course of such a long lifespan, will we design for fluctuations in the earth's climate? In this chapter, we look at climate and at how geologists map its evolution. We identify the variables that influence climate change and examine the current issue of greenhouse gases and "global warming." Armed with an understanding of climate, we conclude the chapter by identifying the design characteristics we must incorporate into our blueprint for reconstruction to allow the earth's rural and urban landscape to interact with a biosphere in evolution.

We begin our look at climate with the basics. What tools do geologists use to piece together the puzzle of the earth's climate, and what do they see when they look back in time?

In this regard, a look at the techniques geologists use to date ancient artifacts and past geological events will be worthwhile. There are two broad categories of dating technologies: *relative* dating methods and *absolute* dating methods.

The dating technologies geologists first pioneered were based on a relative time scale derived from *stratigraphy*, or the study of rock strata. As geological processes lay down layers of sediment, layers of volcanic ash, and layers of other deposits, young strata accumulate on top of old strata. Fossils and artifacts found in a deeper stratum are therefore older than fossils and artifacts found in a stratum closer to the surface. Geologists use this principle to establish the date of artifacts and geological events relative to one another and have correlated strata throughout the world to create an exhaustive record of the earth's past. This was how geologists defined the earth's geological eras, periods, and epochs.

It was not until the advent of absolute dating methods, however, that scientists could anchor their record of geological time to the present and assign artifacts and geological events a calendar date. One of the earliest methods of absolute dating was *dendrochronology*, or tree-ring analysis. The width of growth rings measured in douglas fir and other long-lived species reveals annual growing conditions. By correlating samples from around the world, geologists have created a master index of growth rates that in some areas stretches back four thousand years. *Varve analysis* is a similar technique. In this method, geologists study annual sediment deposits, or varves, laid down in still bodies of water. By counting and correlating varves, they have determined the age of geological events as far back as the glacial retreat at the end of the Pleistocene, ten thousand years ago.

The most important absolute dating techniques are radiometric methods. These include carbon-14, potassium-argon, rubidium-strontium, thorium-230, ionium-thorium, fission-track, and various lead methods. Though each method has different applications, all are based on the principle that radioactive elements decay into more stable forms at a constant rate. Take carbon-14. There are two common carbon isotopes, carbon-12, which is stable, and carbon-14, which decays into carbon-12 and has a half-life of 5,780 years. Metabolic activity maintains carbon-14 in an organism's tissues at the level found in the earth's oceans and atmosphere, which evidence such as cross-dating tells us has been constant within the accuracy and time-frame of the radio-carbon dating technique.[1] On the organism's death, however, metabolic activity no longer replenishes carbon-14, and what is left in an organism's tissue decays into carbon-12. By measuring the ratio of carbon-12 to carbon-14, scientists can date samples as far back as fifty thousand years.

Other radiometric dating methods have other uses. Potassium-argon dates rocks rich in micas, feldspars, and hornblendes. Rubidium-strontium dates igneous and metamorphic rocks. Thorium-230 dates oceanic sediments beyond the range of radiocarbon dating. Ionium-thorium dates deep-sea sediments as far back as 300 thousand years. Fission-track dates micas, tektites, and meteorites between 40 thousand and 1 million years

---

1. Slight variations in atmospheric carbon have been observed and—by correlating with tree ring, ice core, and other data—are used to determining past levels of solar energy output.

old, and lead decay methods date samples to the early Precambrian and the earth's formation.

With an arsenal of dating methods and geological principles at their disposal, how do geologists apply the tools of their trade to unravel the mysteries of the earth's climate?

Of particular importance to geologists are core samples taken from the ocean floors. When microscopic plants and animals that live near the ocean's surface die, their shell remains settle on the floor and accumulate in layers. When studied in a core sample, the relative scarcity or abundance of shells from warm or cold-water species provides a history of environmental change. In addition, the ratio between two oxygen isotopes, oxygen-16 and oxygen-18, found in shell deposits provides a measure of the water temperature in which the organism grew. When we correlate core samples from around the world, the ocean floors provide a detailed climate record that in some instances extends as far back as five million years.

On land, tree-ring and varve analysis are important techniques used to determine past climatic trends. Another important land-based source of climate information comes from core samples taken from the polar ice caps. The thickness and density of ice layers is a measure of annual snowfall, which is a measure of temperature and precipitation. In addition, air trapped in the ice contains pollen and reveals levels of atmospheric gases. Ice cores taken from the Greenland ice sheet allow reasonably dating and analysis as far back as about 100 thousand years. Ice cores taken from the deeper Antarctic ice sheets allow dating and analysis somewhat further back in time.

By analyzing the land and ocean based evidence of the past, geologists have made a rich and detailed record of the earth's geological history—the uplift of mountains, the flooding of continents, the opening and closing of oceans, and the changes in rainfall and temperature that accompanied these events.

What ocean and atmospheric conditions, then, did life in earlier periods experience? Interestingly, the earth's average surface temperature has remained remarkably constant since the advent of life—about 20° Celsius, or 70° Fahrenheit. Though, on the average, temperature fluctuations cancel out, there are ups-and-downs. In general, the earth has experienced relatively long periods of warm weather punctuated by relatively short periods of cold weather. These cold periods, or glacial episodes, occur every 150 or so million years and last a few million years.

In geologically recent times, the earth experienced a glacial episode about 600 million years ago. Another ice age gripped the planet about 435 million years ago. The longest of the earth's ancient ice ages was the *Permo-Carboniferous* glacial period, which began about 300 million years ago. The earth's youngest glacial episode, the *Quaternary*, began about 2.5 million years ago. Although continental ice sheets withdrew from Europe and North America about 13 thousand years ago, most geologists think the earth is still in the Quaternary ice age.

The reason for this is that our climate not only experiences cycles of long warm periods broken by ice ages but fluctuations within ice ages. Each ice age has glacial advances and retreats, more than a dozen documented in the Quaternary. Within these glacial cycles are yet other climatic fluctuations, and within these still others. Prior to the Egyptian Empire, the earth was warmer and wetter than today, the Sahara Desert much like the savannah of East Africa. During the Early Middle Ages, temperatures were cooler than today, a factor believed to have contributed to the turmoil of that period. During the High Middle Ages temperatures were warmer. In Great Britain, grapes grew several hundred miles further north than they do today, a situation that heightened tension with the French. In the 1500s, the world plunged into the *Little Ice Age*.[2] Glaciers overran towns in the Alps. Farmers shifted from grain to potato based agriculture. Vikings abandoned farms and settlements on once arable shores in Greenland. In his march against Russia, Napoleon confronted one of the most brutal winters on record. In the mid-1800s, temperatures began to warm, a trend that with ups-and-downs continued to about 1940. Between that date and 1970 temperatures remained constant and possibly declined. In 1980 they began to increase, to peak in 1998 and remain constant and, by some interpretations, slightly decline since. Even in our lifetime, we experience cycles of rain and drought and of warm and cold temperatures. In a glacial period, there is no such thing as a "normal" climate.

All things considered, one point stands out about the earth's climate and its change over time. Our environment is not and has never been

---

2. What is popularly referred to as the Little Ice Age defines a period when there was a great deal of climatic fluctuation. For this reason, the dates at which the Little Ice Age began and ended are open to interpretation. The period between 1500 and 1850 is widely accepted. Beginning dates, however, range as far back as the 1100s. Many climatologists also consider the Little Ice Age to have ended somewhat earlier, late 1600s to early 1700s.

static, or in a sustained state of equilibrium. The earth's climate has and will continue to be in a state of flux. It is also clear that this change takes place in cycles and that successively smaller cycles are imposed on successively larger cycles—cycles within cycles within cycles. We come to one other conclusion about our climate, and to us it will be no surprise. Geological evidence tells us that climate change does not occur linearly or gradually but in jumps and spurts, across thresholds between the relatively stable states that define overall cycles. The transition between two climatic directions—a glacial retreat and a glacial advance for example—can occur rapidly.

Given the dynamic nature of the earth's climate, what factors account for its change? As one would expect, there is no consensus as to the causes of global climate fluctuation. On one point, however, scientists agree. No single factor accounts for climate change. The earth's climatic cycles are determined by a set of factors, each with its own subset of variables and feedback mechanisms between variables.

One such factor and subset of variables involves the position of the solar system in the Milky Way. Our galaxy completes one rotation every 300 million years. As with tidal processes, two phases appear to exist for this cycle and to coincide with 150 million year glacial events. As the solar system crosses phases, it experiences an abrupt change in galactic environment—different levels of interstellar dust, different magnetic and gravitational fields. These factors impose different stresses within and between the sun and planets. These stresses affect the energy output of the sun and the earth's ability to retain and radiate that energy.

Activity in the solar system is also a factor in climate change, and may be particularly important in determining fluctuations within the 150 million year glacial cycle. The sun has cycles of high and low energy output, including a 22-year sunspot cycle, and cycles within cycles within cycles of high and low energy output. During the height of the Little Ice Age, the measures of sunspot and solar activity available at the time suggest that the sun's output was low with respect to its present level of activity, a period known as the *Maunder Minimum*. Moreover, the earth's orbit around the sun is not a perfect circle but elliptical with an eccentric variation on a 93 thousand year cycle. In addition, the earth's equator tilts with respect to the planet's orbital plane on a 41 thousand year cycle, and the earth experiences a precession cycle, or a wobble, that lasts 26 thousand years. The earth's orbital characteristics vary the amount of energy that reaches the earth from the sun and the distribution of solar radiation across the planet's surface. They also correlate with the quaternary's glacial advances and retreats and are considered to be the dominant factors that account for these events.

The issue becomes more complex when we take into consideration the set of earth-based processes we would expect to factor into climate change. Continental drift affects winds and currents; and the earth is not a perfect sphere but slightly pear shaped with a bulge at its equator. This bulge appears to grow and shrink over time, which may affect energy flow between cold and warm regions. Scientists have postulated ocean salinity as a variable. Fluctuations in and the periodic pole-reversal of the earth's magnetic field, which affect the amount of solar energy the earth reflects into space, are also variables. In addition, the earth experiences cycles of volcanic activity and fault line movement that correspond to cycles of increased and decreased energy transfer from the interior to the surface. There are also incidental factors associated with this release of energy. The uplift of mountains and the spreading of oceans alter winds and currents. Volcanic activity alters atmospheric gases and particulates, which, as the eruption of Krakatau in 1883 and the documented cold winters that followed illustrates, can reflect solar energy into space.

Feedback mechanisms between variables must also be considered. Increased glaciation increases the planet's *albedo*, or its ability to reflect solar energy into space, which may increase cooling, glaciation, and energy reflectivity, perpetuating the cycle and catalyzing a period of cooling and ice buildup. In contrast, a warming trend may increase the level of water vapor in the atmosphere, which as a greenhouse gas may hold in heat, increase warming, and further increase the level of water vapor in the atmosphere.

The last class of earth-based variables is the most complex—the biosphere itself. The average temperature of our planet has remained remarkably constant since the advent of life, and the creative activity of the biosphere has played a clear role in maintaining this temperature, a theory postulated by the British chemist James Lovelock in the 1970s and popularly known as the Gaia Hypothesis. As the earth's temperature changes, forests expand and contract, and algae concentrations in the world's oceans increase and decrease. In ways we have scarcely begun to understand, the creative power of the biosphere embodies the feedback mechanisms to buffer energy flows in a way that maintains the environmental conditions on which life depends.

On one level, climate change is a simple matter. When energy on the earth's surface increases things heat up. When energy decreases things cool down. If, however, we incorporate into our model fluctuations in energy sources, fluctuations in the planet's ability to harness and retain energy, feedback mechanisms between variables, and the biosphere's ability to buffer these and other fluctuations, we must look with awe at the complexity of our environment. Our climate and its ability to change demonstrate the mathematical certitude and creative freedom we can only account for through the dynamics of the creative process. It reveals the behavior we would expect as the universe progressed through its era of life, and as today evolution's trailing arrow reshapes in support of humankind's advance into the age of fulfillment.

This brings us to a contemporary topic, the political and environmental issue of "global warming." As with our discussion on hydrogen as a fuel, the subject is worth a look; because, like hydrogen, it is widely misunderstood. It is with some reluctance, that I delve into the matter, as the political cause of anthropogenic global warming will have run its course before humanity as a whole has made the conscious decision to perfect life on earth. It is a topic illustrative of the misuse of science and the corruption of science by politics, however, and for this reason is of value.

We all have a tendency to reduce complex issues to simple cause and effect models. Environment selects for physical traits and by doing so causes evolution. An asteroid struck the earth sixty-five million years ago and caused the extinction of the dinosaur. The reduction of things to cause and effect interaction is one way we make sense of our world, but it can lead to oversimplification. In actuality, a cause and effect relationship is quite difficult to establish.

With regard to the theory of global warming as described in the popular media and by activists and politicians who have taken on the cause with religious zeal, energy constantly reaches the earth from the sun, and the earth constantly radiates a portion of that energy into space. As long as a balance is struck between energy inflows and outflows, the earth's temperature remains constant. But in experimental situations, certain gases have been shown to act like the glass on a greenhouse. They let more energy in than they let out.

Though rarely cited by global warming proponents, the most potent greenhouse gas is water vapor—which varies in concentration from a fraction of a percent of atmospheric gases at cold temperatures to at saturation occupy about two percent at warm temperatures. Under a clear sky, scorching daytime temperatures in the desert may plunge to below freezing after sunset. Under a cloudy sky, the variation may be tens of degrees less and take place slowly. In tropical areas and areas such as the United States east coast during the summer, humidity levels out day and nighttime temperatures and those without air conditioning suffer through hot, steamy nights. Other than water vapor, there are two primary greenhouse gases: carbon dioxide, which makes up about 0.03 percent of atmospheric gases and is released in metabolism and from the burning of fossil fuels, and methane, which makes up about 0.00015 percent of atmospheric gases and is synthesized in any number of biological processes but is often cited by environmental and animal rights groups as produced in the digestive tracts of cattle. According to global warming theory, our industrial and agricultural practices release greenhouse gases, primarily carbon dioxide, into the atmosphere, which trap heat and increase the earth's temperature.

Though controversial, there is evidence to support this theory. Measurements suggest that atmospheric levels of methane and carbon dioxide and, with ups-and-downs, the earth's average temperature have increased over the last fifty years. But widespread global metrological monitoring only began during World War II, and even the best measurements are open to interpretation. Pundits and politicians casually toss about the figure for the earth's average temperature, but in reality it is quite difficult to determine. Effects such as temperature increases due to pavement and urban development near measuring stations and variations in the calibration of measuring instruments prior to World War II must be taken into consideration. Most of the planet is covered by water, and water has a high specific heat. Ocean temperatures must be considered and, due to ocean depth and currents, measurements may be uncertain. To a degree, past atmospheric gas concentrations can be determined from ice-core samples. Claims such as the recent "carbon dioxide level today is the highest in 20 million years," well beyond the 150 thousand or so years that scientists can extrapolate from Antarctic ice core samples, must be taken with a high degree of skepticism. Statistics are also easily manipulated. Not often pointed out by "global warming" advocates, for example, the earth's temperature began to rise at the end of the Little Ice Age, otherwise we would still be in the Little Ice Age, a trend that began more than one hundred years before widespread industrialization. So difficult is it to determine the earth's average temperature with the accuracy needed to measure annual fluctuations, that many scientists reject figures for this purpose prior to the widespread use of meteorological satellites during the 1970s.

The Quaternary's glacial advances and retreats do appear to coincide with decreases and increases in atmospheric carbon dioxide, such as carbon dioxide levels can be determined, but there is no agreement as to cause and effect. Given the correlation between glacial events and the planet's orbital characteristics along with the preponderance of the greenhouse effect attributable to water vapor, many climatologists feel that increases and decreases in carbon dioxide are the result of and not the trigger for climate change. Warmer temperatures, for example, may increase vegetation, which removes carbon from the atmosphere, but increase the size and frequency of forest fires, the extent of rot and decay, and the release and combustion of methane, all of which add carbon to the atmosphere and in net can lead to a carbon increase.

The most often cited evidence for global warming comes from climatological models. An explosion of computing power the last couple of decades has prompted researchers to attempt to quantify the variables associated with climate change and develop models that predict future rainfall, temperature, and other weather and climatic conditions. These models embody a remarkable degree of analytical engineering. But, like any abstract edifice, they rest on assumptions. What was the sun's energy output at a given time in its past? How do biological processes buffer the earth's climate? Do methane and carbon dioxide have the same greenhouse effect in the atmosphere as in the lab? Is the earth's magnetic field presently in a state of flux? In that the most pervasive greenhouse gas is water vapor, patterns of rainfall and cloud cover are important. Our meteorological models can at best predict a storm a few days out. Are we asking too much of our scientists to predict patterns of rainfall and cloud cover decades into the future? With one set of variables and assumptions, a model tells us the earth is warming. With another, it tells us the earth is cooling. Some say we should prepare for global warming, others for an ice age—a view widely held by the environmental community in the 1970s and embraced by some today. In 2006, Democratic Party Chairman

Howard Dean stated the belief that human activities cause global warming, and global warming will cause the earth to enter an ice age.

Also controversial in global warming theory is the effect of a temperature increase on our lives and environment. Global warming's leading advocate, former United States Vice President Al Gore, recently told us that if we do not drastically cut greenhouse gas emissions in the next ten years it will be too late and the consequences for humanity and the earth will be dire. The issue as to how anyone could come up with a figure like Gore's ten-year window aside, is a warmer climate necessarily a bad thing? The High Middle Ages was a warmer period than today, and it marked one of history's most dramatic stages of advance. Barring massive increases in sea level and other catastrophic changes, which are highly speculative—Venice, which dates to the fifth century, was not under water during the High Middle ages—certain regions would face changes we would consider detrimental and others would face changes we would considered beneficial.

Methane, carbon dioxide, and other greenhouse gases may affect the balance between the energy the earth receives from the sun and the energy the earth radiates into space. Greenhouse gases may play a role in feedback mechanisms and the biosphere's ability to buffer climate change. But the only thing we can say for sure is that greenhouse gases and their emission would be one of many factors. They represent one set of variables and fluctuations imposed on successively larger sets of variables and fluctuations. In part, we have latched onto the cause and effect model of greenhouse gases and global warming because it helps us come to terms with the climatic cycles we experience. In part, we have accepted global warming because it serves our political agendas.

Case in illustration, the first major international agreement to limit greenhouse gas emissions, the politically divisive *Kyoto Protocol*—or to be accurate the Kyoto Protocol to the United Nations Framework Convention on Climate change in December of 1997, named after the Japanese city in which it was proposed.

Established on the premise that greenhouse gases emitted by human activity are causing significant and disastrous changes in the earth's climate, the Kyoto Protocol calls for signatory nations to, by between 2008 and 2012, reduce their emission of methane, carbon dioxide, and four other greenhouse gases to a level that is 5.2 percent below a 1990 baseline. Methods to accomplish this objective called for in the treaty include trapping greenhouse gases by planting trees and reducing greenhouse gases by buying and selling emissions rights, thus placing a monetary value on emissions—carbon offsets.

Some tout the Kyoto Protocol as the most significant environmental treaty ever negotiated. Others feel that it would have little if any effect on global warming. In 2004, the European Environment Agency reported that the fifteen European Union signatory nations were not expected to meet their mandated emissions targets, and United Kingdom Prime Minister Tony Blair questioned the treaty's viability. China, which recently overtook the United States as the world's largest emitter of greenhouse gases, India, Brazil, and other less-developed nations are not required by the protocol to restrict their emissions. In terms of greenhouse gases and global warming, the treaty addresses so little practical as opposed to legal measures—cellulosic ethanol and nuclear power rather than carbon offsets and the establishment of markets to trade carbon offsets—that many feel it has little if anything to do with the environment.

Such is how politicians in the United States accessed it. By design, the Kyoto Protocol places the burden for reducing emissions on countries that have a growing population and that in some way are not exempt. Between 1990 and 2025, European nations had or are expected to have little if any population growth. During that same period, the United States had or is expected to have a population increase in excess of forty percent, due almost entirely to immigration. As such, European nations would have to make few changes to meet the Kyoto emission goals— which they have not been able to do—and the United States would have to decrease per capita emissions, which equates to a decrease in per capita fossil fuel consumption and a corresponding decrease in economic activity, in excess of thirty percent. Depending on the assumptions one makes about immigration, economic growth, and living standards, and if one assumes no massive deployment of nuclear power, this would equate to an economic downturn that would dwarf any post World War II recession. So dire would the consequences be for the United States that President George W. Bush dismissed the treaty out of hand, and the Senate, including legislators who criticized Bush for not signing on, opposed it with a unanimous vote.

The Kyoto Protocol singled out the United States economy to such an extent, that many see the treaty, its recent variations, including that signed on by United States president Barack Obama during the 2009

Copenhagen Convention, and the issue of "global warming," as reflective of an ideology adopted to further a larger political agenda. The more radical members of the environmental community perceive the United States, and the capitalistic ideals it represents, as an obstacle to the socialized, ecologically based society we described with regard to United Nations Agenda 21 guidelines for sustainable development. Prior to the fall of the Soviet Union, the socialist movement propagandized an image of the ideal society that centered on egalitarianism—a worker based utopia. Today, the socialist movement has adopted a new image of the ideal society, an environmentally based utopia. French Prime Minister Jacques Chirac declared Kyoto to be the first step to a world government.

Global warming symbolizes the conflict between economic and environmental ideologies that grips our planet. Even if human released greenhouse gases did play a role in climate change, would it matter? Today, there is little we can do to reduce our consumption of fossil fuels without dramatically reducing living standards in the industrial nations and dramatically limiting progress in the developing nations,[3] a change that when it comes down to it most of us will not accept or impose on others; and, in the future, the consumption of fossil fuels will not be an issue. Our blueprint for reconstruction has solved the problem of greenhouse gases whether or not it existed.

In this regard, to design for the earth's changing climate, we must accept a simple fact. The global climate is in a state of flux. We may be moving into a warmer period. Then again, the glacial retreats we observe today— taking into consideration that we have yet to determine what constitutes a glacial retreat and what constitutes a cycle of glacial variation—may be the fluctuation that thrusts the earth into another major glacial episode. We must accept the earth's climate for what it is, a dynamic expression of creative activity, and engineer for its ups-and-downs.

This we will do in two ways: First, we will design roads and buildings to withstand climate change. We will build the foundation structures of our urban communities high enough to allow for moderate rises in sea level. We will locate roads and cities far enough from rivers to allow for changes in flow and channels. Second, we must humble ourselves and understand that there are climate changes we cannot engineer against.

3. As discussed in chapter 9, *Energy*, the only practical way to at present provide for growth without increasing greenhouse gas emissions is by the deployment of nuclear power.

No farm is productive without water. No city can withstand the advance of a continental glacier. Our blueprint for reconstruction must be flexible. It must allow us to move farms and communities when rainfall patterns change and deserts grow and shrink. Our goal is to take command of evolution's trailing arrow, but we must do so within the biosphere's climatological limits. To design for a changing climate, we must accept the creative nature of climate—in no respect has the earth ever been in a sustained state of climatological equilibrium—and incorporate this understanding into our ideal of earthly perfection.

As we dream of the future and refine our plans for earthly perfection. As we gather and organize the material and human resources we need to begin this most noble of human undertakings, we will set forth on the evolutionary road we, our children and grandchildren, and their children and grandchildren will follow. We will overlay a new landscape on the old, modifying and reestablishing ecosystems, tearing down and rebuilding cities. We will solve humanity's energy problem and put to rest the economic, political, and climatological issues it has imposed. We will accept the creative nature of the earth's climate and design for its dynamic nature.

With this background established, it is time to move into the book's final section and look beyond the early issues and concerns of our task to reconstruct the earth's urban and ecological infrastructure. What, over the decades and centuries, will our blueprint for earthly perfection allow us to achieve?

PART FOUR

Perfection

11

Seed of Genesis

W E BEGAN THIS BOOK with a chapter titled *Perfection*. The universe is in the act of self-perfection—or, more accurately, the universe is the act of self-perfection. The drive for perfection is the sole energy of nature, the unifying force of physics. It is the fuel that motivates creativity, the power that energizes evolution. Just as the universe is driven by the need for perfection, humanity is driven by the need for perfection. The individual seeks self-perfection. Humanity seeks perfection of life on earth. As the book progressed, we developed a vision of earthly perfection and outlined a blueprint to implement our vision. In this section, we look at how this implementation will take place and at how our engineering and architectural ideals will transform our landscape on a global scale. We conclude the section by aligning our ideal of perfection with the larger evolution of consciousness view and the turning point in evolution to which the task of earthly transformation will carry humanity.

Will the earth's first new city be New Los Angeles, New Mexico City, New Buenos Aires? Will it be New Tokyo, New Peking, New Sydney? Will it be New Berlin, New Geneva, New Warsaw? Will it be New Nairobi, New Islamabad, New Johannesburg? Will it be New Moscow, New Vladivostok, New Saint Petersburg? Will it be New Cairo, New Rome, New Paris, New Athens, New Venice, New London? Twelve thousand years ago in Mesopotamia, our ancestors left their nomadic way of life and built the first settlements. What will New Baghdad look like, New Damascus? Wherever we choose to begin, how will we break ground? After we have broken ground, how will our construction efforts proceed and to what end will they climax?

Let us take the hypothetical geological and ecological region we discussed in the book's last section, a division with clear natural boundaries—shores, mountains, and watersheds. For this region, we have a working drawing that tells us where we will establish wilderness areas and draw lines between wilderness categories. It tells us where we will bound intermediate zones, where we will harvest resources, where we will build recreational facilities, and where we will stretch our rural highway system. It tells us where we will place farms and where we will build rural communities and urban centers. In addition, we have mapped soil types, geological faults, and rivers and floodplains. We know the depth of sediment, the location and extent of mineral deposits, and the region's rainfall and drainage patterns. We have also identified human issues concerning the existing landscape—old roads, towns, and cities, old state and private property rights, and old land-use planning bureaucracy.

The members of the economics of fulfillment community in which we belong and through which we will conduct our building efforts have basic material needs. Our first job is to direct our construction efforts to address these material requirements. In this way, we integrate our building efforts with our economic activity and bring about the economic growth and stability we need to support further building efforts. We begin by bringing farmland under the economics of fulfillment blanket. In accordance with the dynamics of an economics of fulfillment system, farmers will acquire land from corporate, government, and other concerns and redraw property lines to meet the farmer's creative needs and to accommodate the nature of the crops.

As we reestablish our connection to the land and rise above ideas of profit and central control, private property ownership, and the rights and responsibilities it entails, will take on a new vigor. In contrast with socialistic ideals of egalitarianism and government ownership of land and resources, private property ownership is intrinsic to economics of fulfillment and is essential for earthly perfection. Only when we take as our own our home and our place on earth, do we have the emotional bond we need to reshape our surroundings to our ideal of perfection. Moreover, as we commence reconstruction, we will not only see our house and farm as our own, we will see our community, city, state, and nation as our own. We will feel a patriotism, one achieved not by virtue of our unity in opposition to an enemy but by virtue of our unity in the understanding of humanity's role and purpose in evolution. We will be proud of what we as individuals and as a community and society achieve. Under economics of fulfillment, private property will take on a significance that is spoken of and longed for but rarely achieved in the present socialist-capitalist economic construct.

On these lines, and in support of our reinvention of agriculture, we will form the social and governing structures that allow us to coordinate our efforts and to advance the art of crop production. These will include businesses, farming alliances, research facilities, and educational programs, and we will use these structures to integrate crop production with hog, dairy, cattle, and poultry production and with the use of organic waste. Free of the drive for profit, we can also responsibly harness the benefits of bioengineering. Today, we use bioengineering as a tool to maximize immediate yields and short-term monetary gains. In the future, we will use bioengineering as a tool to control plant and animal evolution in a way that better integrates artificial ecological systems with one another and with natural ecological systems and that assures stability and productivity. Aquaculture will also become a more important industry in the future and will provide us with the flexibility to manage the ocean's fisheries and ecological systems for sustainable harvests. When patent rights and bottom line figures no longer matter, we can align our creative energies and employ technology to manage and reshape our geological and ecological region as a whole.

In the process of bringing agriculture under the economics of fulfillment blanket, we will build the necessary physical infrastructure. Our first construction projects will be small: homes, barns, dairies, lodges, resorts, and other recreational facilities. Under the economics of fulfillment blanket, however, the companies that build these structures will not operate like today's construction firms.

In the socialist-capitalist framework, the construction process is highly compartmentalized. The architect is responsible for a building's look and style. The engineer is responsible for a building's structural integrity. A general contractor interprets the blueprints and specifications drawn by the architect and engineer and assigns tasks to the various trades that perform the work: framers, roofers, plumbers, landscapers, electricians, bricklayers, and cabinetmakers. For each task there is a profit margin, and the engineer, architect, and general contractor must oversee workmanship to assure that subcontractors do not cut corners to maximize output and minimize cost and job time.

In the future, there will not be such extreme divisions of labor, in particular for small structures. In the economics of fulfillment framework, the builder's objective is not to output a marketable product with a high profit margin but to fulfill his or her creative needs through the construction process. To a greater extent than today, individuals, couples, families, and small companies will design, engineer, and build our homes. The house will be their artistic and engineering achievement, the medium that brings fulfillment into their lives.

There will come a time, however, when we must move beyond homes and small structures and align our efforts to accomplish larger undertakings. We will construct the steel mills, the sand and gravel quarries, and the cement processing facilities that will allow us to lay the roads of our rural highway system and to begin reconstruction of the countryside in a grand way.

Concurrent with the building of our rural highway system, we will construct the first rural communities. Many rural communities will rise on undeveloped land. Many will be built over existing towns. We will demolish streets and buildings and erect coherent villages where we can walk to where we need to go—to where we meet and talk. In addition, we will preserve what is of value from the past. Not every old building is of worth, but there are architectural traditions that define evolution and that we must retain. Moreover, in building our rural communities, we will construct the parking and service facilities that support our rural highway system and that allow the rural community to function as the link to the city. The rural community will be the point where we park our cars and walk to the transit terminals that whisk us to urban centers.

The construction of these structures will be our greatest undertaking. Waves of heavy equipment will demolish block after block of old buildings, and waves of earth moving machines will excavate the site of our first foundation structure. With the groundwork for this structure in place, armies of builders and lines of yet other machines will come in and place and weld steel reinforcement. Still more builders and equipment will transport, place, and cool the tremendous volumes of concrete we will poor as our foundation structure rises to its design height.

With the completion of our first foundation structure, the machines and engineers that built it will move on to the next, and we will erect the urban community itself. Lower levels will rise on the foundation structure; and, in these levels, we will build the transit terminals that link rural communities, industrial sites, and the neighboring communities in our city unit. Towers will reach into the sky. Businesses will open, and we will move into new homes. We will all in some way contribute to the building effort, and many of us will finish our own business and living spaces. As we have described, no one more than the owner takes pride in his or her workmanship. We will also reclaim land built over by the urban sprawl of the past. We will tear apart roads and subdivisions and transform the urban landscape of old into parks, forests, farmland, and rural communities.

Our first rural communities and our first city will reflect our freedom and the unity made possible by our individuality. They will be the physical manifestation brought into being by economic groups that function as islands of abundance and individual expression that have risen out of the waters of socialist-capitalist incoherence.

In time, however, the isolation of these economic units will end. As economic communities grow and merge, we will expand our plans for reconstruction. The rebuilding of our first ecological region will be the training ground for a vast mobilization. Those of us who pioneered its fabrication will lead reconstruction on a global scale. We will identify geological and ecological zones around the world and draft a working drawing for each—a master plan for the earth itself.

In our global blueprint for reconstruction, we will not only deal with the dynamics of our engineering. We will deal with the social ramifications of our engineering. As human beings, it is our nature to seek a better life for ourselves and for our families. Just as today people in Mexico, Central America, and South America are drawn to the United States, often putting their lives at risk in the hope for a better future, people from around the world will be drawn to the economics of fulfillment dream and to the landscape they see taking shape around them. As architects of reconstruction, we must minimize the need for populations to relocate and, when such is not possible, engineer circumstances to eliminate the breakup of families and the destruction of social structure relocation brings. To build New San Diego, New Los Angeles, and New San Francisco, we must build New Mexico City and New Guatemala City. To build New Calgary, New Montreal, and New Vancouver, we must build New Chicago and New Seattle.

As we implement our plan for earthly reconstruction, the socialist-environmentalist ideal of a better tomorrow embodied in notions of the "new world order" and United Nations Agenda 21 guidelines for sustainable development,[1] and that is proclaimed by some in academia and the environmental movement as the "correct" way for humankind to live, will fall from our thoughts. To further the abandonment of this view, we must put behind us the issue at the core of present collectivist thought, the foundational assumption on which the environmental fundamentalism of today is based—the overpopulation of the earth. To make sense of the many "facts" about global population and population growth we come across every day, a look at population demographics is in order.

According to the March 2004 bulletin of the *Population Reference Bureau*,[2] an organization that has been tracking population demographics since 1929, the earth at the start of the twentieth century had a population of about 1.6 billion people. By the end of the century, that number had reached 6.1 billion, with most of this growth taking place after 1960. In line with United Nations projections, the bureau estimates a global population of about seven billion by 2015.

As one would expect, two basic factors influence population dynamics: death and birth. Interestingly, scientists do not attribute the rapid growth in the world's population during the latter half of the twentieth century to an increase in the birthrate. They attribute it to a lowering of the death rate brought on by an increase in life expectancy. In 1950, an individual in the undeveloped world lived about 41 years and in the developed world about 66 years. By the year 2000, these figures had increased to respectively 63 and 76 years.

This stretching-out of lifespan gave rise to what those in population circles call a *demographic transition*. Throughout history, human population has grown rather slowly. Women had a lot of children, but high birth rates were offset by high death rates from war, disease, poor nutrition, and poor sanitation. Modern medicine and agriculture and, above all, extensive sewage and water treatment lengthened the average lifespan. Birthrates exceeded death rates to such an extent that the earth's population exploded. But this does not mean that the earth's population will always grow. Life expectancy must at some point level off; and, as we see in the United States, where by some measures the number has slightly declined, may have reached a plateau at between 76 and 78 years. As life

---

1. See chapter 4, *Roundness of the Earth*.
2. "Transitions in World Population," 3.

expectancy peaks and the death rate catches up to offset the birthrate, population growth will level off.

This takes us to the opposite side of the population equation. Though the primary variable in today's demographic transition is the death rate, the birthrate also comes into play. In contrast, however, the worldwide birthrate, or fertility, the last couple of centuries has plunged.

In the early 1800s, the average American woman gave birth to seven children. By the early 1900s, this number had decreased to four children. Today, the average American woman gives birth to about two children, and the birthrate in Italy, Spain, the Czech Republic and other European nations has dropped to as low as 1.3 children. Even in the developing world, women are opting to have smaller families. In the 1950s, the average woman in Kenya gave birth to 6.7 children. Today she gives birth to 5.2 children. In Asia over the same period, the number has decreased from 5.9 to 2.6 children, in Latin America and the Caribbean from 5.9 to 2.7 children.

Declining fertility has many causes. In the developed world, it is the result of delayed marriage, higher divorce rates, and women choosing to attend college and to opt for career over or in addition to family. As significant, living standards for many working families in the United States and other developed nations have declined over the last decades. Couples that want large families cannot afford them, or have to spend so much time on the job and away from the home that they cannot properly raise them, and, responsibly, have opted for fewer children.

In the undeveloped world, a parallel set of dynamics has come into play. When countries modernize and women become better educated and adopt new roles and lifestyles, families get smaller. In Mali, women with no education have an average of 7.1 children. Women with secondary or higher education have an average of 4.1 children. In Peru, women with no education have an average of 5.1 children. Women with secondary education have an average of 2.4 children and women with higher education, 1.8 children.

Periodically, the *United Nations Population Division* publishes population projections through the year 2050.[3] In a high growth scenario, which is essentially a linear projection of the current growth rate, the earth will be inhabited by 10.6 billion people in 2050. In a more realis-

3. "World Population Prospects 2002 Revision," 1.

tic medium growth scenario, it will be inhabited by 8.9 billion. In a low growth scenario, which better takes into account declining fertility, population growth will peak at about 7.6 billion in 2037 and decline to about 7.4 billion in 2050. Over the years, the Population Division has revised their estimates downward as fertility and birthrates have become better understood. This suggests that the earth's population could reach its maximum sooner than the 2037 estimate. When, in the not-too-distant future, lifespan in the developed world peaks and lifespan in the undeveloped world catches up, the earth's population will begin what some predict will be a sharp decline to at some point settle into a neutral growth pattern where birth and death rates are equal. The issue of overpopulation at the heart of environmental activism is not an issue at all.

Though the assumption of overpopulation on which environmental fundamentalism is based is false, it carries our discussion to a related topic. We must incorporate into our blueprint for reconstruction the means to align our economic and engineering ideals with geopolitics and with the transformation of geopolitics economics of fulfillment and urban and ecological rebirth will make possible. With working drawings and material and human resources at the ready, we can do what in recent years political institutions have failed to accomplish, rebuild a nation devastated by war. Military intervention by the United States and other developed nations may change the course of internal hostilities in a developing nation or depose a tyrannical regime; but, as we have seen time-and-time-again, the economic and political structures we put in place rarely stabilize a nation, and peacekeeping forces must remain. Not since the rebuilding of Japan and Europe after World War II—and, to a lesser extent, the rebuilding of the Eastern Bloc after the fall of the Soviet Union—has the socialist-capitalist economic framework brought forth the political will and the material resources to rebuild nations. We, however, face no limits to reconstruction: no cost and budget constraints, no political and ideological restrictions. In what better way can we overcome ancient ethnic and religious tensions, and their economic catalysts, than to give people the means to work and to express their creativity—the freedom for the human community to align toward the greater cause of global reconstruction?

The non-collective global society of the future rests on the foundation of its internal structures. Business and administrative organizations will form to support global reconstruction. City and regional governments will form where construction is complete—democracy not as we practice it today but as we will practice it under economics of fulfillment, democracy where the freedom of the many is realized through the freedom of the individual.[4] Names and borders may be different, but the state and nation will be stronger and more coherent than they are today. In the urban and ecological fabric we create, however, trade, travel, and communication across borders will be unrestricted. From the country, we will drive to the rural community where we will catch the transit to the city. From the city, we will catch ground transit to nearby cities and air transit to cities further away—no tariffs, no passports, no inspections, no border guards. We have integrated our rural and urban environment with the social structure of the human community. By embracing geography, we free ourselves from geography and open ourselves to the earth.

The years that lie immediately before us will be a time when we are exposed to new ideas and dream of the possibilities. When new ideas take hold, change will follow. Sooner then we may think, the first economics of fulfillment groups will organize, and we will take the first steps to transform our dream to rebuild the earth's urban and ecological infrastructure into physical reality. As was ancient Rome to its citizens, the earth will be an ideal, a statement, an expression of who we are as human beings. It is the place that, when we ask ourselves what kind of a world we want to live in, we long to call home. The years will pass. Occupied in reengineering the planet, memories of war, poverty, and pollution will dim, and the earth's population will stabilize at the level needed to support reconstruction. The decades will pass. New cities will have arisen around the globe, and we will have forgotten life without meaning. Just as today it is scarcely possible to imagine consciousness without the capacity of reflective thought and learning, in the future it will be scarcely possible to imagine consciousness without the internalization of the universe, its evolution, and its immediate and ultimate future. The centuries will pass. We will learn as we go, refining our ideal of perfection, coaxing the earth to an ever more inspiring form.

4. See Book 2, *Economics of Fulfillment*, chapter 10.

12

A Life

W<small>E HAVE DRAFTED A</small> design for the earth of tomorrow, a vision of the rural and urban landscape we have the creative power to make reality. Yet, our blueprint for reconstruction is just that, a plan, a drawing. Before we move on and present the book's conclusion, we will attempt to give our plan a touch of color and humanity. We have a general understanding of what the rural and urban landscape of tomorrow will be like, but we do not know what our life may be like in that landscape. It is time to give our imagination a chance to run.

Our story begins on a fall day, high on the flanks of an urban tower. A gentle breeze crosses a granite terrace, which overlooks the urban communities in the first city unit of New Los Angeles. On the terrace stands a thirty-nine year old woman. Her name is Tanesia. With her are her mother and father, her husband, and her fifteen-year-old daughter. Tanesia isn't a leader in the reconstruction movement. She has no training as an engineer and has drawn no city plans and designed no skyscrapers. Her skill is as an ironworker. Yet, Tanesia looks on the city structure that rises before her with the greatest sense of accomplishment. In her way, she made it possible. As Tanesia gazes on the towers that soar around her, and on the transit lines that run between urban communities and that give the city unit the look of a blossoming flower, she also feels a sadness. As grand as the city structure may be, it is complete. The job is over. There will be no more iron to sling.

The completion of the city unit marked the end of a period in Tanesia's life that began three thousand miles away and almost forty years ago. Tanesia was born in Detroit, where until the age of five she lived in an apartment with her mother and father.

Tanesia was too young to remember, but her mother had told her that even when she was two and three years old she loved to draw and to build things. Most children do Tanesia felt when she was old enough to think about these things; but, in her mother's eye, she had a special knack for it, a special love. Her father thought so too. She was a builder, he had told her. She was someone who liked to feel things with her hands and to shape them to her liking.

What Tanesia clearly remembered, though, was that when she was almost old enough to start school she and her mother drove her father to the bus station so that he could catch a coach to Los Angeles. There were great things happening on the West Coast, he had told them. He would go there and see for himself and, if everything looked safe and all right, find a job and a place to live so that Tanesia and her mother could join him.

Her father called when he arrived in Los Angeles to let them know he got there okay, and he called a month later and told them to come. Tanesia and her mother packed what they could fit into their car and gave the rest of their things to a man who fixed the car so that it wouldn't break down on the trip. Tanesia was too young to worry about such concerns, but she was excited about traveling such a long way and she could feel that her mother was too.

The government in Washington D.C. didn't approve of what was going on in California and was stopping people from driving that far west, so they headed south through Mexico and arrived in Los Angeles a week later to find the city no different than they had seen on television—smog, traffic, and urban unrest. Tanesia's father was waiting for them at the house where they would live, and he told them not to be scared and that things weren't as they appeared. For them, life would be much better here, he said. The stores were stocked. They didn't have to pay rent. The schools were good, and the clinics didn't cost anything.

Tanesia didn't know if the schools were any better than those in Detroit because she hadn't gone to school at their old home, but she liked school. Her teachers recognized her talent for drawing and for working with her hands and encouraged her to pursue these interests. Her mother also went to school, and so did her father. But it never occurred to Tanesia that it was an odd thing for parents to do. The parents of most of the kids in her class went to school. It was just something parents did.

In Tanesia's class, the teachers helped her learn to read books with pictures of art and buildings and to learn to calculate and to draw on paper and computers. They also taught her to make things from clay and wood and to think of new things to make and new ways to make old things. School was exciting for Tanesia. There was always something to do or to read or to think about, and there was always time to play and for her parents to visit.

As Tanesia's mother explained, school for adults was different. Her mother's school was a place where grownups could learn what they were good at and what they liked to do. For a long time, Tanesia's mother didn't know what she was good at or liked to do. Then some people came to her school to talk about a farm they were building on the ocean, and she decided to study the sea and its life. While she was learning about these things, Tanesia's mother talked to the people who were building the ocean farm and told them that she wanted to be part of what they were doing. She said that she would open a fish market in the city so they could take turns working on the farm and in town and not get bored and be able to spend more time with their families. The fish farm people liked the idea, and they agreed to work together.

Unlike her mother, Tanesia's father knew right away what he wanted to study in school. He liked to work with wood and in school learned to design cabinets and furniture and to use the tools that made these things. After he knew enough to get started, he found a building near their home and filled it with all kinds of power tools. Tanesia loved this building and its tools. When she was old enough, her father let her help him draw boards through their blades and to clean up piles of sawdust.

In Detroit, her parents had always talked about money. Tanesia didn't understand why, but money wasn't a problem in Los Angeles. They had the things they needed, and her parents could go to school and learn to do what they liked. And, it seemed, they always had time to spend with her. After a while, no one talked about money, and Tanesia forgot all about it.

As the years past, Tanesia's father worked in the furniture shop, and her mother worked on the ocean farm and in the fish stand she had opened in the city. Tanesia spent time in school but also worked with her mother in the market and traveled with her on the boat to the ocean farm where they would feed, harvest, and gut the fish. But her favorite thing was building cabinets and furniture with her father. She was good at calculations and drawing plans, but mostly she liked working with the tools—cutting and joining the wood.

As a teenager, school was less important to Tanesia than when she was younger, but she would take classes when she had a need to learn something or when she wanted to be with her friends. One of her interests was welding, and it was in a class she had taken to learn to weld steel beams that her life changed. She met a man. He was her age and shared her love of tools and building. The two became close, and when Tanesia turned nineteen they married.

Tanesia and her husband lived in the house with Tanesia's parents. For a time, the young couple worked with Tanesia's mother and father in the fish market and woodworking shop. Then Tanesia took a job as a metalworker in a shipyard that scraped old oil tankers. Her husband went to school to train as an engineer and then went to work in the steel mill that processed the iron from the shipyard into steel beams and reinforcement rods. Work for both of them took a backseat when, at the age of twenty-four, Tanesia gave birth to her daughter.

Tanesia settled into family life, and so did her husband and parents. As a mother, Tanesia spent most of her time at home, where she taught her daughter to draw and to build with blocks. Her husband and parents also spent time with the child, who was the center of all of their lives. As a family, they provided Tanesia's daughter with the time and attention she needed and each other with the time to keep up with their work and interests.

Tanesia had a good life, but she never took time to think about it. Her parents had told her stories about the way things were before they moved west, but it was another time and their struggle to make ends meet and to deal with the difficulties of their old city didn't make much sense to Tanesia. She was happy with the life she knew, and she was never bored. Things were happening all around them—great things. The television was showing plans and models of a new city to be built on a part of Los Angeles periodically torn up by riots and where few people now lived. The new city was to be enormous, with soaring towers and with urban communities linked by transit lines to form city units, which were surrounded by lakes and farms. Everything about the new city was grand and magnificent, fanciful and extraordinary. And they were told that in only a few years they would live there.

Taken by the plans and pictures of the new city, and by the machines that would build it, Tanesia felt a renewed yearning to work with her hands. She spent more time in her father's shop and, when her daughter was old enough to start school, left her job at the shipyard and went to work as a welder building the cranes that would be used to construct the new city. At the age of thirty-two, Tanesia, her husband, her daughter, and her mother and father gathered around the television to watch lines of bulldozers tear through block after block of old brick and wooden buildings and lines of dump trucks haul away the debris.

Never had Tanesia felt such excitement. Never had she seen such machines as those that were shaping the land on which the new city would rise. It was all she, her husband, and her parents talked about, the new city. It was all her daughter and her daughter's friends talked about, how when they grew up they were going to build new cities all over the world. It was as if the entire universe had focused on what was happening in their lives. Everyone felt a sense of purpose, a feeling that they were part of some great mission to transform the earth. This, Tanesia thought, must have been what it felt like when, generations ago, people united and took up arms to fight for freedom.

Only their cause was one of peace and renewal. Giant excavation machines dug the enormous hole in which the foundation structure of the first urban community would be built, and Tanesia left her job building cranes to join one of the hundreds of crews that would lay and weld the steel reinforcement and place the steel molds in which the foundation concrete would be poured. Crews and machines worked day and night until, six months later, the foundation structure was in place. Some crews moved on to build the next foundation structure. Tanesia and her team of ironworkers stayed to raise the steel beams of the first urban community. They began with the vast expanse of interior and exterior space on the community's lower levels and then went on to build the transit terminals and the first great towers.

Everywhere in Los Angeles, activity was at a frenzy. Lines of construction equipment rolled along freeways. Steel mills spewed out truckload after truckload of girders and reinforcement. Cement factories and sand and gravel quarries fed plants that mixed concrete and that transported it into the new city on conveyers and in enormous buckets suspended from cables. Rail lines brought food from the countryside. Ships unloaded slabs of granite from Canada and slabs of marble from China and Brazil. Inspired by the structure taking shape before them, and by the excitement of its builders, people throughout the old city were tossing away their dollars, embracing economics of fulfillment, and joining the reconstruction movement. New schools were opening to provide the education people needed to work on the construction crews and to open businesses in the new city.

When they finished raising the girders of the tower where Tanesia and her family would live, she took time off her job and worked with her mother, father, and daughter to finish her father's new woodworking shop, which was located on one of their urban community's lower levels near her mother's new fish market. After this was done, and the walls and stonework of the tower were in place, Tanesia and her family set to work finishing the three thousand square foot space that would be their home. All around them, people were doing the same. Restaurants and businesses were opening, and people were leaving their old homes to move into the urban community. As people vacated buildings in the old city, bulldozers cleared away streets, freeways, and structures to make way for the other urban communities in the city unit.

After Tanesia and her family had moved into their tower home, she went back to work with her crew of ironworkers and went on to build the next urban community in the city unit and then the one after that. Tanesia's mother stayed busy in the fish market and on the ocean farm, and her father stayed busy building furniture for the new homes. Tanesia's husband continued to work at the steel mill, which extruded many of the beams and reinforcement bars Tanesia's crew raised and welded into place.

Tanesia's daughter who, while helping to finish their tower home had taken a liking to tile work, taught families how to set marble and granite floor tiles in their homes. She also had an aptitude for mathematics and dreamed of one day designing a city. Then again, she liked to cook and wanted to open a seafood restaurant next to her grandmother's fish market. She also liked to swim, paint, sculpt, and grow flowers and wanted to show her art along the urban community's walkways, landscape its terraces, and coach children at one of their community's pools. Such was as it should be, Tanesia thought. Her daughter could do it all.

As work progressed on the other urban communities in the city unit, life settled down for Tanesia and her family. The transit made it easy to get around, and they didn't have to go into the old city and deal with its problems. Tanesia divided her time between her family and a new job training ironworkers. This didn't give her a chance to work with her hands, so she also built furniture in her father's shop. Things were also settling down in their city unit. Hundreds of businesses were open, and work on the lakes and parkland around the unit was underway. Tanesia and her family spent time in their tower home, but the community was also their home, their extended family and place to belong.

Seven years after they had broken ground on Los Angeles's first urban community, their city unit was declared complete. Tanesia and her family gathered on the granite terrace of their tower home to look upon their accomplishment. Their city unit was home to a half million people and more inspiring than anything they had imagined when years ago they had studied its drawings and models. As she gazed over the cluster of urban communities that rose before her, Tanesia's feelings were of awe and pride, of satisfaction in the work she had done. They were also of sadness, for the city unit was finished. The job was done.

The completion of the city unit marked the end of a period in Tanesia's life that began when as a child she and her parents had traveled west. Looking back, Tanesia felt a nostalgia at the great venture that she had in part pioneered. But her feelings for the past didn't last for long. Her city unit may be done, but in the distance, bulldozers were tearing apart roads and freeways to build other city units. New Los Angeles would be home to twenty million people; and, after it was complete, there would be other cities to raise and more iron to sling. Tanesia was a builder who worked with her hands, and she had a world to shape.

13

Fulfillment

A S TEMPTING AS IT may be to lose ourselves in our dream of tomorrow and the world we have within us the capacity to build, we cannot forget that we and all that we create are part of a greater evolution. The
universe advances through our internalization of an ever refining vision
of the future. In the most outward way, this vision will manifest through
the reshaping of our rural and urban landscape into ever more functional
and aesthetically satisfying forms. The endeavor of earthly perfection will
carry us from decade to decade, from generation to generation, from century to century. It will also carry us to an evolutionary crossroads. Our
quest for earthly perfection can have only one outcome, and it will open
the door to an even more inspiring future.

In our evolution of consciousness vision of creation, the universe advances through the creative process. Life and humanity progress through
a mechanism of creative cycles and the growth and collapse of uncertainty
that culminates in the crossing of thresholds to higher stages of existence
and forward movement. In the process of this evolution—by way of trial
and error and groping to create greater consciousness and fulfillment—
creativity has two directions. The leading arrow of the creative process
drives evolution forward—to states of greater autonomy, complexity, and
consciousness. The trailing arrow reshapes and discards prior evolutionary forms in support of the universe's overall progress. The creative process builds on and creatively discards the old to create the new.

By way of this mechanism, the universe advanced through its periods
of emergence, structure, life, and understanding. From the threshold to
reflection, some one hundred thousand years ago, to the present, we have
sought to comprehend the nature and purpose of our existence. Today,
we have reached the turning point in evolution that marks the first stage
of our achievement of this goal—the initial phase of our transcendence

to a new level of awareness. Today, we rise above knowledge and create that to which knowledge must lead—wisdom. We see the universe as an evolution of consciousness and by doing so internalize its origin, nature, and purpose. We build within ourselves an awareness of the past, present, and future—a sense of knowing as much a part of our being as our ability to think reflectively, as much a part of who we are as our awareness of consciousness. Today, the universe crosses the threshold to meaning. Today, evolution marks the turning point that lifts humanity beyond the age of understanding and into the universe's final period of advance—the age of fulfillment.

Empowered by meaning, we look back on our past and, armed with the wisdom it embodies, set forth to create our future. We rise above socialism and capitalism and embrace a new economic philosophy—an economics for the universe's concluding period of movement—an economics of fulfillment. Economics of fulfillment is the tool that frees us from economic burden, the mechanism that lifts us above notions of scarcity, competition, winners and losers, and government control of human economic and other activities. It is also the tool that frees us from the conflict between economic and environmental interests that today grips the planet. The dogma of profit and exploitation drives us blindly forward in a quest to harness the material resources of the earth. The dogma of environmentalism drives us blindly back in a quest to restore the biosphere to some past evolutionary level, to some imagined state of environmental correctness. This collision between ideologies polarizes the management of our countryside and imposes on our cities an urban plan that lingers from the days of horse-and-buggy transportation.

We are the conduit through which the biosphere reinvents itself. The human role is not to exploit the earth. Neither is it to return the earth to a past evolutionary state. We are not stewards of the environmental status quo. We are commanders of evolution's trailing arrow. As creative beings, we have the right, the means, and the duty to mold our world in support of humankind's advance and the universe's becoming. Our position as overseers of evolution's trailing arrow empowers us to rise above the dogmas of old and perfect life on earth. We envision the rural and urban landscape we want to live in and draft a blueprint to shape that landscape.

Fulfillment

> The objective of our blueprint for reconstruction is to establish the urban and ecological infrastructure that allows the individual to exercise his or her creative power and that through the process of planning and construction serves as an avenue for the individual's creative expression.

The universe advances through the evolution of the human community, and the human community advances through the evolution of the individual human being. Our task is to engineer a rural and urban landscape that provides each of us creative freedom and that allows each of us to express our creativity through the task of earthly perfection. The act of creation is as essential to our well-being as is perfection itself.

In the countryside, we will have true wilderness, those few remaining large tracts of land in Alaska, Siberia, Australia, the Canadian Northwest territories, and other remote parts of the world. We will also have wilderness managed for access, areas where wildlife does not pose a danger to hikers and backpackers and where trails are mapped and patrolled. Bordering wilderness areas will be intermediate zones. In these areas, our goal is to stabilize ecological systems at a level of complexity that provides for recreation and for the sustainable harvest of timber, minerals, and other resources. Nestled within intermediate zones will be agricultural land. Here our objective is to maximize output and advance the art of food production by embracing the tool of properly administered science and by addressing the creative needs of the farmer. Throughout the countryside there will be farms, homes, rural communities, and a highway system that ties the post-reflective landscape into a cohesive unit.

The building block of the future city will be the urban community. It is the geographical area where most of us will choose to live and work—our home, the corner of the world where we feel that we belong. The urban community will be large enough to support a diverse economy and to sustain the more intimate, nestled, and complex levels of social organization we will create in the future. It will be small enough to allow us get to where we need to go on foot. The engineering groundwork of the urban community will be the foundation structure. It is a reinforced concrete edifice designed to withstand the worst natural hazard. In this structure, we will locate our community's mechanical systems. On this structure, we will build the urban community itself.

Building the foundation structure will be an exercise in standardization, mass production, and engineering creativity. Building the urban community will be an exercise in architectural creativity and artistic inspiration. Sweeping lower levels will house shops, schools, businesses, restaurants, medical facilities, and transit terminals. Soaring towers will house still more businesses—and living spaces made grand by the inspiration of ownership. A transit system will link urban communities to form city units and link city units to create successively larger urban structures, a nestled and coherent design able to accommodate a large population on a small geographical area.

As a design concept, our blueprint for reconstruction shows us that we can overlay a functional urban and ecological infrastructure on the earth's surface, one where we integrate human activities into the biosphere and by doing so manage evolution's trailing arrow to meet our needs as evolving beings in an evolving universe. Like a living organism, tomorrow's rural and urban landscape will consist of centers of human activity nestled within centers of human activity, each level integrated by a coherent communication and transportation system.

As we dream of the future and draft the working drawing for earthly perfection. As we gather and organize the material and human resources we need to begin this most noble of undertakings, we will set forth on the evolutionary road we and generations for centuries to come will follow. We will overlay a new landscape on the old, modifying and establishing ecosystems, tearing down and rebuilding roads and cities. Through our perfection of the landscape, we will solve humankind's energy problem and learn to accommodate a climate that by its dynamic and creative nature is in flux. To connect with our journey through time and its destination, we will retain the elements of our voyage's unfolding, those historical, architectural, and other traditions that define human progress and that jar in our evolutionary memory the recollection of humankind's achievement.

But our task of urban and ecological reconstruction is not without end. Our quest for perfection can have only one outcome, only one resolution—perfection. In the distant future, the day will come when we will have attained the earthly ideals to which we have aspired. Through the trial and error of the creative process, we will have reshaped our rural and urban landscape to achieve such a degree of beauty and functionality that we will be content with our surroundings. We will have sculpted our world to such a level of achievement, to such a level of sophistication, to such a level of artistic and engineering expression that there will be nothing left to do—nothing more to dream, nothing more to build, nothing more to preserve. With the achievement of earthly perfection, humankind will reach yet another turning point in its evolution to greater consciousness and fulfillment.

Evolution's leading arrow exists only in the present. At the moment we perfect life on earth, we exhaust the evolutionary potential of reconstruction. Our time on earth and all we have created is, as it can only be, a step along the way. From the highest tower of all we have dreamed and accomplished, we will gaze into the future. Driven by the vision before us, we will gather the strength to abandon what, for the breadth of the human experience, we have strived to create, and take the first step on a still more inspiring evolutionary road. Humankind will extend its presence into the furthest reaches of the universe's age of fulfillment. With these words, we put behind us our blueprint for reconstruction and align with the conclusion we reached in *Threshold to Meaning: Book 1, Evolution of Consciousness*.[1] With the achievement of earthly perfection, humankind will transcend evolution's final turning point on the relic of past creation we call the earth. Before our eyes—brilliant and drawing us toward it—we will face the future that all we have learned and built made it possible for us to create. Our perfection of life on earth has prepared us to move on. Humankind will look back with wonder on all it has achieved in the physical realm of trailing-edge evolution, direct its gaze forward in time, and reach to bring forth the universe's ultimate state of being.

1. See *Book 1, Evolution of Consciousness*, chapter 14.

14

Mission for Completion

As I HAVE WRITTEN in this and in the earlier books in the *Threshold to Meaning* series, it is my belief that the universe is today crossing an evolutionary threshold and that this transcendence is taking place within us—the universe's threshold to meaning. This transformation represents the most profound leap in human consciousness since the emergence of reflective thought and learning more than one hundred thousand years ago. As an outcome of the rebirth of consciousness now taking place within us, humanity will move beyond socialism and capitalism and embrace the economics of fulfillment philosophy. Empowered by economic freedom, we will set forth to reshape our surroundings—to perfect life on earth. At the center of life's reinvention is the individual. Earthly perfection is our personal quest to grow and learn, our personal drive to express our character and to refine and idealize the world around us. As an individual, how do we find our place in earthly perfection? What is our mission for completion?

The world to which we aspire stands before us. With our acceptance of the universe as an evolution of consciousness, a vision of the earth's future landscape revels itself as if it had always been there and only awaited our ability to understanding it. It is not a socialist-environmentalist utopia, where the needs of the individual are subverted for the collective good and the good of the planet as it exists in the imagination of a ruling elite. It is a flexible and evolving design driven by the individual and the individual's quest for freedom and opportunity. In our vision of tomorrow, we will have wilderness areas, and we will have intermediate zones where we can extract resources. In our vision, we will have farms and rural communities, and we will overlay our countryside with a coherent and maintainable highway system. In our vision, we will have cities, and we will engineer our cities to support our evolving personal and social needs. Meaning opens the way to economics of fulfillment, and economics of fulfillment opens the way to reconstruction of the earth's urban and ecological infrastructure.

Yet, to take command of evolution's trailing arrow and engineer our surroundings to meet our needs as evolving beings in an evolving universe, we must put the present behind us. In today's world, the human experience is shaped by a scarcity-based interpretation of existence. From this dogma emerges conflict between economic and environmental interests and the urban sprawl and incoherent communication and transportation systems that have resulted. Contemporary economic and environmental doctrine fails to embrace the creative nature of the human being and the evolving nature of the biosphere. The world of today offers a way of life gripped by an antiquated, materialistic understanding of the universe, but one in which we have grown to accept.

Caught in the uncertainty of our time, we struggle to find our way. We know where we are and where we want to be, but we do not know how to get there. Our future is clear, but the road to that future awaits our invention. What role do we as an individual play in this transition? What can we as a unique man or woman do to thrust humankind across the threshold to the future and the better way of life it promises?

As do all roads of substance, our path to earthly perfection begins within ourselves. Our first step is to open ourselves to the possibility that we can play a role in the future, that we can bring about urban and ecological reconstruction on a global scale. Global perfection is achieved through the perfection of that within reach of the individual, by our own hands. What tasks do we find rewarding? What social relationships do we find fulfilling? How, in a personal way, do we envision the future of the earth's urban and ecological infrastructure? As a human being—as one who is essential to the human experience—what is our ideal of perfection?

The universe exists at the moment of transformation between two evolutionary eras: an age of money and uncertainty and an age of wisdom and human creativity. Empowered by our individuality and by the unity it allows us to bring forth, our lives align with the lives of others on the same path. As we accept and undertake this journey, we recognize our individual course to perfection and direct our creative energy toward its resolution.

Through each of us, the universe advances. Within each of us lies the universe's origin. Within each of us lies the universe's periods of emergence, structure, life, and understanding. Within each of us lies the universe's threshold to reflection and the potential to cross the universe's threshold to meaning. Within each of us lies the capacity to—with all the exuberance and all the humbleness the journey has instilled—set forth into the universe's final evolutionary era, the age of fulfillment. On birth, we became part of the great movement that is the universe, and our existence will carry us to the end in which it must climax. When we each in our own way arrive, we and the universe will achieve the enduring perfection for which we yearn and have for millennia strived to create.

Bibliography

"Analysis of the Lieberman-Warner Climate Security Act (s.2191) using the National Energy Modeling System (NEMS/ACCF/NAM)." Washington DC: The National Association of Manufacturers and the American Council for Capital Formation, 2008.

Architectural Theory: From the Renaissance to the Present. London: Taschen, 2006.

Arms, Myron. *Riddle of the Ice: A Scientific Adventure in the Arctic*. New York: Anchor Books, 1998.

Ashby, W. Ross. *An Introduction to Cybernetics*. New York: Methuen and Company, 1964.

Atkins, Stephen S. *Historical Encyclopedia of Atomic Energy*. Westport: Greenwood Press, 2000.

Bacon, Edmund N. *Design of Cities*. New York: Viking Press, 1967.

Balling, Robert C. *The Heated Debate: Greenhouse Predictions Versus Climate Reality*. San Francisco: Pacific Research Institute for Public Policy, 1992.

Barnett, Lincoln. *The Universe and Dr. Einstein*. New York: Time Incorporated, 1962.

Bedogne, Vincent F. *Commonsense Guide to Current Affairs*. Eugene: Resource Publications, 2009.

Bergson, Henri. *Creative Evolution*. Translated by Arthur Mitchell. New York: The Modern Library, 1944.

Bergson, Henri. *The Creative Mind*. Translated by Mabelle L. Andison. New York: Greenwood Press, 1968.

Bergson, Henri. *Time and Free Will: An Essay on the Immediate Data of Consciousness*. Translated by F. L. Pogson. New York: Greenwood Press, 1968.

Bertalanffy, Ludwick Von. *General System Theory: Foundations, Development, Applications*. New York: George Braziller, 1968.

Blatt, Harvey. *Principles of Stratigraphic Analysis*. Boston: Blackwell Scientific, 1991.

Bradley, John. *Learning to Glow: A Nuclear Reader*. Tucson: University of Arizona Press, 2000.

Brown, E. and R. B. Firestone. Table of Radioactive Isotopes. New York: Wiley Interscience, 1986.

Brown, Harry. *The Phoenix Project: Shifting from Oil to Hydrogen*. Phoenix: SPI Publication and Production, 2000.

Brown, Theodore L., et al. *Chemistry: The Central Science, 6th ed.* New Jersey: Prentice Hall, 1994.

Bryant, Edward. *Climate Process and Change*. New York: Cambridge University Press, 1997.

Calvin, William. *The Ascent of Mind: Ice Age Climates and the Evolution of Intelligence*. New York: Bantam, 1990.

Campbell, Bernard G. *Humankind Emerging, 7th ed.* New York: HarperCollins, 1996.

Bibliography

Delfgaauw, Bernard. *Evolution: The Theory of Teilhard de Chardin.* Translated by Hubert Hoskins. New York: Harper and Row, 1969.

Dimensions of Mind. Edited by Sidney Hook. New York: Collier Books, 1961.

Dodson, Edward O. *The Teilhardian Synthesis, Lamarckism & Orthogenesis.* Lewisburg: American Teilhard Association, 1993.

Drake, Francis. *Global Warming: The Science of Climate Change.* New York: Oxford University Press, 2000.

Dubos, René. *Celebrations of Life.* New York: McGraw-Hill, 1981.

Dunn, Seth. *Hydrogen Futures: Toward a Sustainable Energy System.* Washington DC: World Watch Institute, 2001.

Encyclopedia of Climate and Weather. Edited to Stephen Schneider. New York: Oxford University Press, 1996.

Environmental Design: An Introduction for Architects and Engineers. Edited by Randall Thomas. New York: E & FN Spon, 1996.

Farb, Peter. *Humankind: What We Know About Ourselves. Where We Came From and Where We are Headed. Why We Behave the Way we do.* New York: Bantam, 1978.

Flavin, Christopher. *Beyond the Petroleum Age: Designing a Solar Economy.* Washington DC: World Watch Institute, 1990.

Giere, Ronald N. *Understanding scientific reasoning, 3rd ed.* Orlando: Holt, Rinehard, and Winston, 1991.

The Grammar of Architecture. General editor Emily Cole. New York: Barnes & Noble, 2005.

Grenet, Paul. *Teilhard de Chardin: The Man and His Theories.* Translated by R. A. Rudorff. London: Souvenir Press, 1965.

Hedman, Richard. *Fundamentals of Urban Design.* Washington DC: Planners Press, 1984.

Houghton, John. *Global Warming: The Complete Briefing.* New York: Cambridge University Press, 1997.

Hurley, Patrick J. *A Concise Introduction to Logic, 5th ed.* Belmont: Wadsworth Publishing Company, 1994.

Huxley, Julian. *New Bottles for New Wine.* New York: Harper and Brothers, 1957.

Jantsch, Erich. *Design for Evolution: Self-Organization and Planning in the Life of Human Systems.* New York: Braziller, 1975.

Jantsch, Erich. *The Evolutionary Vision: Toward a Unifying Paradigm of Physical, Biological, and Sociocultural Evolution.* Boulder: Westview Press for the American Association for the Advancement of Science, 1981.

Jantsch, Erich. *The Self-Organizing Universe: Scientific and Human Implications of the Emerging Paradigm of Evolution.* New York: Pergamon Press, 1980.

Jones, Steven, R. D. Martin, and David Pilbeam. *The Cambridge Encyclopedia of Human Evolution.* Cambridge: Cambridge University Press, 1992.

King, Ursula. *Christ in All Things: Exploring Spirituality with Teilhard de Chardin.* Maryknoll: Orbis Books, 1997.

Krane, Kenneth. *Modern Physics, 2nd ed.* New York: John Wiley & Sons, 1996.

Lane, David. *The Phenomenon of Teilhard: Prophet for a New Age.* Macon: Mercer University Press, 1996.

Linacre, Edward. *Climate Data and Resources.* New York: Rutledge, 1992.

Marshall, Alex. *How Cities Work: Suburbs, Sprawl, and the Roads Not Taken.* Austin: University of Texas Press, 2000.

Bibliography

Marx, Karl. *The Communist Manifesto*. Chicago: H. Regency Co., 1954.

Mechanics of Intelligence: Ross Ashby's Writings on Cybernetics. Edited by Roger Conant. Seaside: Intersystems Publications, 1981.

Milne, Anthony. *Earth's Changing Climate: The Cosmic Connection*. Garden City: Prism Press, 1984.

Norris, Robert E. and L. Lloyd Haring. *Political Geography*. Columbus: Charles E. Merril and Company, 1980.

Ogden, Joan M. and Robert H. Williams. *Solar Hydrogen: Moving Beyond Fossil Fuels*. Washington DC: World Resources Institute, 1989.

Ortega y Gasset, José. *The Revolt of the Masses*. New York: W. W. Norton and Company, 1960.

Peixoto, José Pinto. *Physics of Climate*. New York: American Institute of Physics, 1992.

Peterson, Willis. *Principles of Economics, 4th ed*. Homewood: Richard D. Irwin, 1980.

Pletsch, Carl. *Young Nietzsche: Becoming a Genius*. New York: The Free Press, 1991.

Prigogine, Ilya and Isabelle Stengers. *Order Out of Chaos*. New York: Bantam, 1984.

Resnick, Robert and David Halliday. *Basic Concepts in Relativity Theory and Early Quantum Mechanics*. Englewood Cliffs: Prentice Hall, 1991.

Rozenzweig, Cynthia. *Climate Change and the Global Harvest: Potential impacts of the Greenhouse Effect on Agriculture*. New York: Oxford University Press, 1998.

Russell, Bertrand. *The ABC of Relativity*. New York: Signet, 1925.

Rykwest, Joseph. *Seduction of Place: The City in the Twenty-First Century*. New York: Pantheon, 2000.

Scargill, David I. *The Form of Cities*. New York: St. Martin's Press, 1979.

Schofield, John. *Cost-Benefit Analysis in Urban and Regional Planning*. Boston: Allen & Unwin, 1987.

Serway, Raymond A. Principles *of Physics*. New York: Harcourt, 1994.

Sheehan, Molly. *Reinventing Cities for People and the Planet*. Washington DC: World Watch Institute, 1999.

Stevens, William K. *The Change in the Weather: People, Weather, and the Science of Climate*. New York: Delacorte Press, 1999.

Stikker, Allerd. *The Transformation Factor: Towards an Ecological Consciousness*. Rockport: Element Books, Inc., 1993.

Teilhard de Chardin, Pierre. *Activation of Energy*. Translated by Rene Hague. London: Collins, 1970.

Teilhard de Chardin, Pierre. *Building the Earth*. Translated by Noel Lindsay. Wilkes-Barre: Dimension Books, 1965.

Teilhard de Chardin, Pierre. *Christianity and Evolution*. Translated by Rene Hague. New York: Harcourt Brace Jovanovich, 1971.

Teilhard de Chardin, Pierre. *Early Man of China*. New York: AMS Press, 1980.

Teilhard de Chardin, Pierre. *Human Energy*. Translated by J. M. Cohen. London: Collins, 1969.

Teilhard de Chardin, Pierre. *Hymn of the Universe*. Translated by Gerald Vann. New York: Harper and Row, 1961.

Teilhard de Chardin, Pierre. *Let Me Explain*. Edited by Jean-Pierre Demoulin. Translated by Rene Hague. London: Collins, 1970.

Teilhard de Chardin, Pierre. *Letters From a Traveller*. New York: Harper and Row, 1962.

Teilhard de Chardin, Pierre. *Letters to Two Friends*. 1926–1952: New York: New American Library, 1968.

Bibliography

Teilhard de Chardin, Pierre. *Man's Place in Nature.* Translated by Rene Hague. New York: Harper and Row, 1966.

Teilhard de Chardin, Pierre. *On Love and Happiness.* San Francisco: Harper and Row, 1984.

Teilhard de Chardin, Pierre. *Science and Christ.* Translated By Rene Hague. New York: Harper and Row, 1968.

Teilhard de Chardin, Pierre. The *Appearance of Man.* Translated by Robert T. Francoeur. New York: Harper and Row, 1965.

Teilhard de Chardin, Pierre. *The Divine Milieu: An Essay on the Interior of Life.* New York: Harper and Row, 1960.

Teilhard de Chardin, Pierre. *The Future of Man.* Translated by Norman Denny. New York: Harper and Row, 1964.

Teilhard de Chardin, Pierre. *The Heart of the Matter.* Translated by Rene Hague. New York: Harcourt Brace Jovanovich, 1979.

Teilhard de Chardin, Pierre. *The letters of Teilhard de Chardin and Lucile Swan.* Edited by Mary W. Gilbert. Washington DC: Georgetown University Press, 1993.

Teilhard de Chardin, Pierre. *The Making of a Mind: Letters from a Soldier-Priest, 1914–1919.* Translated by Rene Hague. New York: Harper and Row, 1965.

Teilhard de Chardin, Pierre. *The Phenomenon of Man.* Translated by Bernard Wall. New York: Harper and Row, 1959.

Teilhard de Chardin, Pierre. *The Vision of the Past.* Translated by J. M. Cohen. London: Collins, 1966.

Teilhard de Chardin, Pierre. *Toward the Future.* Translated By Rene Hague. New York: Harcourt Brace Jovanovich, 1975.

Teilhard de Chardin, Pierre. *Writings in Time of War.* Translated By Rene Hague. New York: Harper and Row, 1968.

"Transitions in World Population." *Population Reference Bureau Population Bulletin 59* (2004). Online: http://www/prb.org.

Weinberg, Alvin. *Nuclear Reactions: Science and Trans-Science.* New York: American Institute of Physics, 1992.

"World Population Prospects the 2002 Revision." New York: The United Nations Department of Economic and Social Affairs Population Division, 2003.

www.ingramcontent.com/pod-product-compliance
Lightning Source LLC
Chambersburg PA
CBHW070919270326
41927CB00011B/2645